Exploring the World of Cereal Crops

Edited by Timothy Tse

Published in London, United Kingdom

Exploring the World of Cereal Crops
http://dx.doi.org/10.5772/10.5772/intechopen.1002163
Edited by Timothy Tse

Contributors
Amol N. Nankar, Anca Corina Fărcaş, Brenda Contreras-Jiménez, Carlos Martín Enríquez-Castro, Charity Pisa, Eduardo Morales-Sánchez, Faith Matiza Ruzengwe, Jana Kholová, Keerthi Chadalawada, Krithika Anbazhagan, Loreen Murimoga, Maha Khalfalla, Maria Simona Chiş, Martin J.T. Reaney, Maya Chandra, Morleen Muteveri, Ngavaite Chigede, Pengfei Zhang, Qi Zhang, Rachel Gwazani, Rashmi Banoriya, Rumbidzai Blessing Nhara, Sivasakthi Kaliamoorthy, Sunita Choudhary, Tharanya Murugesan, Timothy Tse, Venkata Subramanya Vara Prasad Kodukula, Vetriventhan Mani, Vincent Vadez, Zhichao Li, Zoltán Győri

First published in London, United Kingdom, 2025 by IntechOpen
IntechOpen is the global imprint of INTECHOPEN LIMITED, registered in England and Wales, registration number: 11086078, 167-169 Great Portland Street, London, W1W 5PF, United Kingdom

For EU product safety concerns: IN TECH d.o.o., Prolaz Marije Krucifikse Kozulić 3, 51000 Rijeka, Croatia, info@intechopen.com or visit our website at intechopen.com.

British Library Cataloguing-in-Publication Data
A catalogue record for this book is available from the British Library

Exploring the World of Cereal Crops
Edited by Timothy Tse
p. cm.

This title is part of the Agricultural Sciences Book Series, Volume 18
Topic: Agronomy and Horticulture
Series Editor: W. James Grichar
Topic Editor: Ibrahim Kahramanoglu
Associate Topic Editors: Murat Helvaci and Olga Panfilova

Print ISBN 978-0-85466-749-9
Online ISBN 978-0-85466-748-2
eBook (PDF) ISBN 978-0-85466-750-5
ISSN 3029-052X

If disposing of this product, please recycle the paper responsibly.

IntechOpen Book Series

Agricultural Sciences

Volume 18

Aims and Scope of the Series

The importance of agriculture cannot be overstated. It helps sustain life, as it gives us the food we need to survive and provides opportunities for economic well-being. Agriculture helps people prosper around the world and combines the creativity, imagination, and skill involved in planting crops and raising animals with modern production methods and new technologies. This series includes two main topics: Agronomy and Horticulture, and Animal Farming. This series will help readers better understand the intricacies of production agriculture and provide the new knowledge that is required to be successful. The success of a farmer in modern agriculture requires knowledge of events happening locally as well as globally that impact input decisions and ultimately determine net profit.

Meet the Series Editor

W. James Grichar has been employed with Texas A&M AgriLife Research for over 45 years with an emphasis on research in agronomy, plant pathology, and weed science. He obtained his BS from Texas A&M in 1972 and his Masters of Plant Protection in 1975. He has published 195 journal articles, over 330 research reports and briefs, 11 book chapters, and over 300 abstracts of profession meetings. He also directs research in many crops including corn, grain sorghum, peanuts, and sesame. He has held various positions in different professional societies including the American Peanut Research and Education Society, Southern Weed Science Society, and Texas Plant Protection Conference in addition to being Associate Editor for Peanut Science and Weed Technology. Significant accomplishments have included spearheading efforts to determine the optimum planting time for soybean production along the upper Texas Gulf Coast. These efforts have shown growers that soybean yields can be improved by 10 to 20% by following a late March to early April plant date. He also has been instrumental in developing a herbicide program for peanut production in the south Texas growing region. Through the development and use of herbicides that are effective against major weed problems in the south Texas region, peanut yields have increased by 25 to 30%.

Meet the Volume Editor

Dr. Timothy Tse is an accomplished researcher specializing in environmental and agri-food research, with a particular emphasis on sustainability, fermentation technologies, and value-added products. His research contributions have led to over 30 scientific publications and seven book chapters. With extensive experience in analytical methodologies and next-generation sequencing, Dr. Tse has helped to develop advancements in ethanol purification technologies and investigated enhancements of agri-waste byproducts through fermentation to create high-value products (e.g., therapeutic compounds). His research is driven by a commitment to sustainability and innovation, using cutting-edge fermentation processes to foster impactful collaborations and contribute to advancing bio-based research.

Contents

Preface

Cereal crops are the cornerstone of global food security, feeding billions of people while also playing an integral role in agricultural systems. The cereals market gross production value is projected to exceed US$12 billion in 2024. Cereal grains are rich in biological compounds, including vitamins, dietary fibers, and other essential nutrients. These nutrients have been observed to improve human health, such as reducing human diseases. Beyond food, these crops also play a significant role in agri-food and biofuel production, with wheat, maize, and rice being the three predominant cereals supporting the world's population.

As the world's population approaches 9 billion by 2050, addressing the dual challenges of malnutrition and increased demand for cereals becomes critical. This includes efforts to develop drought-tolerant crops and optimize processing technologies that reduce antinutritional factors while enhancing the bioavailability and bioaccessibility of essential nutrients. The impact of climate change on agricultural production and food security further intensifies these challenges, underscoring the importance of strategies like crop rotation, improved fertilization, and pest control. Other research has focused on advancements in breeding to enhance nutritional traits in cereals. In addition, sensor-based technologies have gained attention as a tool for monitoring grain quality related to human nutrition and food safety. Finally, cereal storage remains critical for maintaining the food supply chain and promoting global food safety and security. Innovations in storage, including automation and complex engineering solutions, can monitor key parameters to optimize storage conditions, extend shelf-life, and prevent contamination (e.g., microbial growth).

Altogether, this book will explore the scientific, cultural, and economic aspects of cereal crops, providing insights into innovative practices, emerging trends, and the pathways to ensure the resilience and productivity of these essential food sources.

Timothy Tse
Department of Food and Bioproduct Sciences,
University of Saskatchewan,
Saskatoon, Canada

Chapter 1

Introductory Chapter: Cereal Crops

Timothy Tse and Martin J.T. Reaney

1. Introduction

Cereal crops remain a staple source for nourishment globally. The domestication of rice (*Oryza sativa*) in China 10,000 years ago [1], along with the domestication of maize (*Zea mays*) in Southern Mexico and Central America and wheat (*Triticum aestivum*) in the Fertile Crescent of the Near East [2], has significantly contributed to the transformation of human civilization. The three cereals, wheat, maize, and rice form the dietary foundation for most of the world's population [3]. In addition, with the global population projected to increase by over 9 billion by 2050 and the challenges posed by climate change, there is an urgent need to increase cereal production by approximately 70% [3]. Dietary changes have also been noted as a separate concern from those of climate change, sustainability, and human and animal health [4]. This trend is leading to increased per capita consumption of protein including plant-based proteins [5, 6].

Cereal crops play crucial role in providing essential nutrients for human diets. However, most cereal crops are processed for animal feed, the creation of other food products or components, as well as industrial applications [7]. Cereal grains and whole grain cereal products are rich sources of biologically active compounds such as dietary fibers (e.g., cellulose, lignin and lignans, etc), sterols, tocopherols, tocotrienols, alkylresorcinols, phenolic acids, vitamins, microelements, and fibrous polymers [8]. These compounds have been investigated for value-added applications in other industries, such as in pharmaceutics [9, 10]. The consumption of these nutrients has been shown to reduce heart disease and diabetes [11], improve blood glucose regulation [12] and weight management [13], and lower the risk of certain cancers [13]. Additionally, other bioactive substances present can exhibit antioxidant, anti-inflammatory, and antibacterial effects, aiding in the prevention of human diseases [14, 15].

Ready-to-eat products derived from cereal crops such as granola, breakfast cereals, hot cereals, etc), offer a versatile selection of foods with nutritional value while facilitating storage, and simplifying preparation. These products are formulated from processed grain (e.g., corn, oats, wheat, rice, etc.), legumes, and nuts [16]. They offer nutritional benefits, including low-fat, high-fiber content, which can lead to increased work and social performance, increased longevity, and enhanced overall health [17]. Due to the enhanced protein and nutritional profiles, including carbohydrates, fatty acids, fibers (**Table 1**), ready-to-eat breakfast cereals have become an important dietary component in developing countries, and a nutritional staple for Western breakfasts [20].

In addition to their nutritional components, cereal crops are also rich in lignocellulose content. This makes fractions from these crops suitable as a feedstock for second generation [21] biofuel production [22], in the production of pulp and paper [23], and

Nutrients	Cereal Grains											
	Amaranth	**Barley**	**Buckwheat**	**Cornmeal**	**Millet**	**Oats**	**Quinoa**	**Brown Rice**	**Rye**	**Sorghum**	**Wheat**	**Wild Rice**
Calories (kcal)	167	159	154	163	170	175	166	165	152	148	153	161
Protein (g)	6.10	5.62	5.96	3.65	4.96	7.60	6.35	3.39	4.65	4.78	6.16	6.63
Total lipid (g)	3.16	1.03	1.53	1.62	1.90	3.11	2.73	1.44	0.73	4.56	1.11	0.49
Carbohydrate (g)	29.36	33.07	32.18	34.60	32.78	29.82	28.87	31.31	34.14	32.44	32.01	33.71
Fiber (g)	3.0	7.8	4.5	3.3	3.8	4.8	3.1	1.6	6.8	3.0	4.8	2.8
Calcium (mg)	72	15	8	3	4	24	21	4	11	6	15	9
Iron (mg)	3.42	1.62	0.99	1.55	1.35	2.12	2.06	0.58	1.18	1.51	1.58	0.88
Magnesium (mg)	112	60	104	57	51	80	89	52	50	74	65	80
Phosphorus (mg)	251	119	156	108	128	235	206	140	149	130	229	195
Potassium (mg)	229	203	207	129	88	193	253	112	230	163	194	192
Sodium (mg)	2	5	0	16	2	1	2	2	1	1	1	3
Zing (mg)	1.29	1.25	1.08	0.82	0.76	1.79	1.40	0.96	1.19	0.75	1.87	2.68
Copper (mg)	0.24	0.22	0.50	0.09	0.34	0.28	0.27	0.13	0.17	0.13	0.25	0.24
Manganese (mg)	1.50	0.87	0.59	0.22	0.73	2.21	0.92	1.68	1.16	0.72	1.36	0.60
Selenium (µg)	8.4	17.0	3.7	7.0	1.2	25.4	3.8	10.5	6.3	5.5	40.2	1.3
Vitamin C (mg)	1.90	0.00	0.00	0.00	0.00	0.00	0.00	0.00	0.00	0.00	0.00	0.00
Thiamin (mg)	0.05	0.29	0.05	0.17	0.19	0.34	0.16	0.24	0.14	0.15	0.19	0.05
Riboflavin (mg)	0.09	0.13	0.19	0.09	0.13	0.06	0.14	0.04	0.11	0.04	0.05	0.12
Niacin (mg)	0.42	2.07	3.16	1.63	2.12	0.43	0.68	2.92	1.92	1.66	3.03	3.03
Pantothenic (mg)	0.66	0.13	0.55	0.19	0.38	0.61	0.35	0.67	0.66	0.17	0.42	0.48
Vitamin B6 (mg)	0.27	0.14	0.09	0.14	0.17	0.05	0.22	0.22	0.13	0.20	0.19	0.18

Nutrients	Cereal Grains											
	Amaranth	Barley	Buckwheat	Cornmeal	Millet	Oats	Quinoa	Brown Rice	Rye	Sorghum	Wheat	Wild Rice
Folate (dietary folate equivalents; μg)	37	9	14	11	38	25	83	10	17	9	19	43

Reconstructed from Oldways Whole Grains Council [18] and USDA Nutrient Database [19].

Table 1.
Nutritional profile for select cereal grains, based on 45 g dry weight.

Category	Global Production (Million Tonnes)
Maize	1163.5
Wheat	808.4
Rice	776.4
Barley	155.9
Sorghum (pseudo-cereal)	57.6
Millet	30.9
Oats	26.4
Rye	13.1
Other Cereals	8.71
Buckwheat (pseudo-cereal)	2.24
Quinoa (pseudo-cereal)	0.159

Data taken from the Food And Agriculture Organization of the United Nations, 2024 [26].

Table 2.
Global production (tonnes) of cereal and pseudo-cereal grains in 2022.

as a packaging material [24]. They are also used as a source of flour that can be used as a resource for manufacturing sticking pastes, and biofuels (e.g., bioethanol) [22]. Therefore, cereal crops play a major role in the evolution of agri-food systems, driving towards sustainable growth and production.

As the global population continues to rise, the production of cereal crops is also increasing. The cereals market gross production value is projected to reach US$12.12 bn in 2024, with an annual growth rate of 3.07% between 2024 and 2028 [25]. Global production of cereal and pseudo-cereal grains from 2022 is detailed in **Table 2**.

With increased production, various cereal storage methods have been exploerd to maintain product quality from storage to consumption [27]. The five primary storage methods are bulk storage, underground pit storage, bag storage, shed storage, and silo storage, with silos being the preferred method [27]. Factors such as moisture content and poor ventilation can degrade the quality of cereal products during storage. However, advancements and innovations in aeration, refrigerated storage, modified atmospheric storage and hermetic storage systems [27] have led to the development of new technologies to improve storage conditions and extend product quality before processing and downstream applications. This book will also discuss advancements in cereal storage systems.

2. Conclusion

In conclusion, the world of cereal crops presents a comprehensive and intricate network that is crucial to global food security, nutrition, and agricultural economies. From staple grains that are the cornerstone of diets worldwide to lesser-known pseudo-cereals gaining recognition for their nutritional benefits and sustainability, cereals offer a rich and vital diversity. This book delves into the complexities and nuances of cereal storage and production, exploring the challenges and opportunities that lie ahead in an era marked by changing climates, evolving dietary preferences, and the relentless pursuit of sustainability. Through an interdisciplinary lens, we will examine the scientific, cultural, and economic aspects of cereal crops, offering insights into innovative practices, emerging trends, and the potential pathways to ensure the resilience and productivity of these essential food sources. By focusing on both the traditional staples like wheat, rice, and maize, as well as the promising potential of pseudo cereals, this book endeavors to provide a comprehensive overview that enlightens, informs, and inspires a deeper appreciation for the cereals that sustain the world.

Author details

Timothy Tse* and Martin J.T. Reaney*
University of Saskatchewan, Saskatoon, Canada

*Address all correspondence to: timothy.tse@usask.ca and martin.reaney@usask.ca

References

[1] Molina J, Sikora M, Garud N, Flowers JM, Rubinstein S, Reynolds A, et al. Molecular evidence for a single evolutionary origin of domesticated rice. The Proceedings of the National Academy of Sciences. 2011;**108**(20):8351-8356. DOI: 10.1073/pnas.1104686108

[2] Gustafson P, Raskina O, Ma X, Nevo E. Wheat evolution, domestication, and improvement. In: Carver B, editor. Wheat Science and Trade. Ames, IA: Wiley-Blackwell; 2009

[3] Neupane D, Adhikari P, Bhattarai D, Rana B, Ahmed Z, Sharma U, et al. Does climate change affect the yield of the top three cereals and food security in the world? Earth. 2022;**3**:45-71. DOI: 10.3390/earth3010004

[4] Sá AGA, Moreno YMF, Carciofi BAM, Carciofi. Plant proteins as high-quality nutritional source for human diet. Trends in Food Science & Technology. 2020;**97**:170-184. DOI: 10.1016/j.tifs.2020.01.011

[5] Orellana-Palacios JC, Hadidi M, Boudechiche MY, Ortega MLS, Gonzalez-Serrano DJ, Moreno A, et al. Extraction optimization, functional and thermal properties of protein from cherimoya seed as an unexploited by-product. Food. 2022;**11**(22):3694. DOI: 10.3390/foods11223694

[6] Özdemir EE, Görgüç A, Gençdağ E, Yılmaz FM. Physicochemical, functional and emulsifying properties of plant protein powder from industrial sesame processing waste as affected by spray and freeze drying. LWT. 2022;**154**(January):112646. DOI: 10.1016/j.lwt.2021:112646

[7] Batey IL. The diversity of uses for cereal grains. In: Cereal Grains. San Francisco, CA, US: Woodhead Publishing; 2010. pp. 45-46. DOI: 10.1533/9781845699529.1.45

[8] Bartlomiej S, Justyna R-K, Ewa N. Bioactive compounds in cereal grains – Occurence, structure, technologic significance and nutritional benefits – A review. Food Science and Technology International. 2021;**18**(6):559-568. DOI: 10.1177/1082013211433079

[9] Mohamed HI, Fawzi EM, Basit A, Kaleemullah L, R., Sofy, M. R. Sorghum: Nutritional factors, bioactive compounds, pharmaceutical and application in food systems: A review. Phyton-International Journal of Experimental Botany. 2022;**91**(7):1303-1325. DOI: 10.31604/phyton.2022.020642

[10] Hadidi M, Garcia SR, Ziogkas D, McClements DJ, Moreno A. Cereal bran proteins: Recent advances in extraction, properties, and applications. Critical Reviews in Food Science and Nutrition. 2023:1-25. DOI: 10.1080/10408398.2023.2226730

[11] Kissock K, Neale E, Beck E. Whole grain food definition effects on determining associations of whole grain intake and body weight changes: A systematic review. Advances in Nutrition. 2021;**12**:693-707

[12] Kelly S, Hartley L, Loveman E, Colquitt J, Jones H, Al-Khudairy L, et al. Whole grain cereals for the primary or secondary prevention of cardiovascular disease. Cochrane Database of Systematic Reviews. 2017;**8**:CD005051. DOI: 10.1002/14651858.CD005051.pub3

[13] Aberg S, Mann J, Neumann S, Ross A, Reynolds A. Whole-graine processing

and glycemic control in type 2 diabetes. A randomized crossover trial. Diabetes Care. 2020;**43**:1717-1723. DOI: 10.2337/dc20-0263

[14] Gaesser G. Whole grains, refined grains, and cancer risk: A systematic review of meta-analyses of observational studies. Nutrients. 2020;**12**:3756. DOI: 10.3390/nu12123756

[15] Tse TJ, Wiens DJ, Chicilo F, Purdy SK, Reaney MJT. Value-added products from ethanol fermentation – A review. Fermentation. 2021;7(4):267. DOI: 10.3390/fermentation7040267

[16] Fast RB. Manufacturing technology of ready-to-eat cereals. In: Fast RB, Caldwell EF, editors. Breakfast Cereals and How They are Made. 2nd ed. St. Paul, Minnesota, USA: American Association of Cereal Chemists; 2020

[17] National Diary Council. Breakfast: Its effects on health and behavior. Dairy Counc Dig. 1993;**64**(2):7

[18] Oldways Whole Grains Council. Compare Nutrients in Various Grains. 2024. Available from: https://www.wholegrainscouncil.org/whole-grains-101/health-studies-health-benefits/compare-nutrients-various-grains [Accessed: February 21, 2024]

[19] United States Department of Agriculture. FoodData Central. Available from: https://fdc.nal.usda.gov/ [Accessed: February 21, 2024]

[20] Jones JM. Cereal nutrition. In: Fast RB, Caldwell EF, editors. Breakfast Cereals and How they Are Made. 2nd ed. St. Paul, Minnesota, USA: American Association of Cereal Chemists; 2020

[21] Tse TJ, Wiens DJ, Reaney MJT. Production of bioethanol – A review of factors affecting ethanol yield. Fermentation. 2021;7(4):268. DOI: 10.3390/fermentation7040268

[22] Kumar SPJ, Kumar NSS, Chintagunta AD. Bioethanol production from cereal crops and lignocelluloses rich agro-residues: Prospects and challenges. SN Applied Sciences. 2020;**2**(10):1673. DOI: 10.1007/s42452-020-03471-x

[23] Chibudike HO, Udohitinah JS. Evaluation of pulp and paper making potentials of cereal straws (agro-wastes). Journal of Engineering Research. 2009;**14**(4):31-36

[24] Vargas F, González Z, Sánchez R, Jiménez L, Rodríguez A. Cellulosic pulps of cereal straws as raw material for the manufacture of ecological packaging. BioResources. 2012;7(3):4161-4170

[25] Statista. Cereals – Canada. 2023. Available from: https://www.statista.com/outlook/io/agriculture/cereals/canada [Accessed: January 10, 2024]

[26] Food and Agriculture Organization of the United Nations. FAOSTAT. Crops and livestock products. Available from: https://www.fao.org/faostat/en/#data/QCL/visualize [Accessed: February 19, 2024]

[27] Pekmez H. Cereal storage techniques: A review. Journal of Agricultural Science and Technology. 2016;**B6**(2):67-71. DOI: 10.17625/2162-6264/2016.02.001

Chapter 2

Processing of Millets

Rumbidzai Blessing Nhara, Charity Pisa, Ngavaite Chigede, Rachel Gwazani, Morleen Muteveri, Loreen Murimoga and Faith Matiza Ruzengwe

Abstract

The necessity for countries in sub-Saharan Africa (SSA) to be self-sustaining in the fight against food and nutrition insecurity is of crucial importance to maintain their autonomy. Promoting indigenous, drought-tolerant crops is a potential way of mitigating the impacts of climate change and supplementing maize, whose productivity has declined due to dependency on erratic rain-fed agriculture. Millets are known for their high amount of macro- and micronutrients (such as B vitamins, potassium, phosphorus, magnesium, iron, zinc, copper and manganese). However, millets also contain significant amounts of anti-nutritional factors (polyphenols, enzyme inhibitors and phytates), resulting in low bioavailability of the minerals and proteins. This has led to employing a number of processing techniques during millet meal production to reduce these effects. Hence, this chapter focuses on evaluating millet processing techniques applied (e.g., soaking, dehulling, steaming, controlled germination and roasting) and their influence on the anti-nutritional factors, nutritional composition and functional properties of millet meals based on the available literature reports. This review demonstrated the importance of millet processing technologies in removing anti-nutritional factors that could reduce the bioavailability or bioaccessibility of essential nutrients.

Keywords: millets, soaking, dehulling, steaming, roasting, controlled germination, nutritional composition, anti-nutritional factors

1. Introduction

Billions of people globally have been shown to be experiencing malnutrition as well as food insecurity [1]. The global target set by the United Nations to end hunger by 2030 is far from being attained [2]. Rapid population growth and climate change, compounded with economic meltdown over the past few decades, have greatly impacted food security [3]. It is thus imperative to transform the current food system to achieve food and nutrition security by providing an affordable, healthy and nutritious diet to all. Millets (sorghum, pearl millet and finger millet) are climate-resilient crops that can be grown with minimum inputs even under unfavourable agricultural conditions and are tolerant to heat, drought and floods [4, 5]. This makes them the crops of choice in light of climate change and limited natural resources because good productivity can be obtained even in marginal areas with water scarcity.

Millets have the potential to play a significant role in the fight against malnutrition and food insecurity in children, adolescents and the community at large, as well as promote health and immunity [6]. They contain an abundant source of essential macro and micronutrients such as carbohydrates, proteins, dietary fibre, lipids and vitamins [7]. Phytochemicals and antioxidant properties in millets give them the ability to prevent the onset of cardiovascular diseases, cancer and diabetes, among others. Additionally, millets are key for people with celiac disease [7].

The production of millet as a source of human food to meet dietary requirements has increased over the past few decades [8]. Despite their benefits, millets are still predominantly consumed by the poor population because of a lack of awareness of processing technologies and appropriate preparation methods. The nutrients, bioactive compounds and functions of millet grains can be influenced by preparation techniques such as decortication/dehulling, soaking, germination/malting, milling, parboiling and fermentation [3, 8]. According to Amadou et al. [9], due to the small grain size, dehulling is not favourable to millets as the process causes nutrient loss. However, dehulling of millets can also improve the bioavailability of nutrients and consumer acceptability of millet-based products. Milling removes the bran and germ layers, causing significant loss of fibre and phytochemicals [9]. Hence, there is a need to select an appropriate method for each millet type during the production process.

Processing millets into various value-added by-products, such as flour, flakes, nutritional bars, cookies, cakes and extruded products, can increase their consumption and utilisation to enhance health, nutrition and food security. The use of millet combined with other food crops to develop novel value-added food alternatives is an emerging and promising area of food technology [6]. Nutrition-rich fermented foods and beverages from millet also form a major part of the diet in most African countries. Fermentation by lactic acid bacteria increases the protein content and improves digestibility [10, 11]. Millet processing technologies afford the opportunity to prepare a wide range of products, hence giving consumers a variety of choices of acceptable, savoury and nutritious food products [12].

Apart from the processing of millet into value-added production for human consumption, it can be processed as a source of livestock feed. Studies have investigated the use of millets in ruminant and non-ruminant diets [13]. For instance, Cisse et al. [14] reported that inclusion of pearl millet of up to 50% in broiler grower and finisher diets had no negative effect on their performance with comparable results to corn on metabolisable energy and digestible amino acids. An enhancement in growth and feed efficiency in broilers was reported after replacing corn with pearl millet [15]. Pearl millet was also reported to be able to fully replace maize in high-supplement diets for confined cattle [13]. Hassan et al. [16] observed that for grazing beef cattle during the dry season, millet processing increases the digestibility of dry matter and dietary nutrients.

Millet's production and processing have drawn the attention of various food security stakeholders worldwide to enhance its better utilisation and reduce hidden hunger in the world. This chapter discusses the processing techniques of millets, the effect of processing on nutrients, anti-nutritional factors, bioavailability and bioaccessibility of nutrients, and the value-added products from millets.

2. Processing techniques

Processing of millet grains can be defined as the conversion of the millet grains into edible forms with an enhanced quality. Various processing techniques are therefore available for the conversion of millet grains into value-added edible products. These processing techniques can be classified into two basic types: hot processes, in which heat is either applied or created during the treatment process and cold processes, in which the temperature of the grain is not increased significantly [17]. The various processing methods are discussed below.

2.1 Soaking

Soaking involves mixing millets in either cold or hot water for a period of 6 to 24 h at various temperatures ranging between 30 and 70°C [18]. To reduce contamination from harmful microorganisms, hot (above 40°C) or cold water, <7°C, can be used. This softens and swells the grain and reduces the cooking time, while also enhancing digestibility [19], palatability and feed value [20, 21]. Portable water is periodically (e.g., 3 h intervals) changed during soaking to a total soaking time not exceeding 24 h. Kate and Singh [22] also reported increased mineral bioavailability and decreased anti-nutritional compounds such as phytic acid. After soaking, the product can then be crushed through crimped rollers to crush the grain. Unfortunately, soaking does exhibit some disadvantages, including the risk of souring during storage. In addition, the increased seed size, or swelling, after soaking can lead to storage capacity limitations [23]. Furthermore, micronutrients such as iron, phosphorus and calcium present in pearl millet have been found to be reduced following soaking in an acidic medium [22].

2.2 Dehulling or dehusking

Dehulling is the process of removing the outer coat of grain, also known as decortication. The hulls (outer covering of the grain) are high in fibre and low in digestibility in both humans and monogastric animals. When these are removed, feed value improves due to increased voluntary feed intake and digestibility [24]. Furthermore, bioavailability of existing nutrients and a decrease in anti-nutrients have been reported [22]. Other nutritive ingredients, such as iron, zinc and calcium, are compromised. Ikwebe et al. [25] reported decreases in nutrient content ranging from 2 to 7% after dehulling of the following: iron (Fe) from 40.3 to 39.3 mg/kg; zinc (Zn) from 38.53 to 35.97 mg/kg and magnesium (Mg) from 311.06 to 306.45 mg/kg. The decrease in these micronutrients suggests that these nutrients are concentrated in the outer coverings of millet grains [26].

2.3 Steaming

The grain is first steam-cooked for 15–30 min to increase the moisture content to 18–20%. The moistened grain is then passed through rollers, yielding a thin, flat flake that is then dried. Steaming has been shown to increase millet polyphenols, which improves antioxidant capacity [24]. This will have a net positive effect on human nutrition [27]. Steaming has been shown to significantly improve the whiteness of millets [28]. In livestock, flaked grain is considered more acceptable and has a slightly higher digestibility value than unprocessed grains. For example, sorghum starch digestibility increased from 42–91% after steaming and flaking [17].

2.4 Controlled germination

This is a natural processing technique that is an extension of the soaking method. The germination period ranges from 48 to 72 h at an ambient temperature of 25 to 30°C. More than 48 h of germination (sprouting) has been shown to result in a loss of grain dry matter with negligible nutritional improvement [22]. Controlled germination enables biological activation of grains, which improves their nutritional and functional properties [23]. Saleh et al. [21] reported an increase in protein digestibility after germination, which was attributed to a reduction in anti-nutritional factors such as phytic acid, tannins and polyphenols, which are known to interact with proteins, forming insoluble complexes that reduce their bioavailability for the body. Controlled germination reduces the component of hydrolysable tannins in sorghum while increasing phosphorus content, improving the grain's feeding value for both humans and animals [29]. Humidity, oxygen and temperature are all critical environmental factors in controlled germination.

2.5 Roasting

Roasting is a form of heat treatment in which the grain is oven-roasted between 120 and 180°C for 3–15 min. It can also be done traditionally with a gas fireplace and an iron pan used as a hotplate that is heated by a flame [30]. It improves the flavour, palatability, voluntary feed intake and feed efficiency of millets [24]. It also destroys heat-sensitive anti-nutritional factors like phytates and tannins [31]. This increases the nutritional value of the final product for both humans and monogastric animals [32]. Roasting at high temperatures around 120°C has been shown to reduce phytates by 34.89% [30]. Wang et al. [24] found that roasting time and temperature affect total polyphenol levels in millets. For ruminant animals, it increases the flow of undegradable digestible dietary flow to post-ruminal sites [17].

2.6 Grinding

This is the process of reducing grain size to a form suitable for the required recipe [22]. A hammer mill can be used for this process. The final grain particle size is determined by the size of the sieve used. The method is simple and inexpensive. It generally improves the digestibility of small hard grains such as sorghum, increasing their feed value. On the other hand, grinding has been shown to significantly reduce the shelf life of millet grains by increasing the free fatty acid (FFA) content, which can cause rancidity in millet flour [22].

Ruminants prefer coarse to fine grinding. Grinding is also crucial in piggery as they are poor chewers of feed. Grinding is essential for maximum performance in poultry kept under intensive conditions as they do not have access to grit, which is used in the gizzard to aid grain breakdown [17]. Excessive grinding should be avoided because it does not improve digestibility or performance and can cause problems with pig health. Fine grinding can produce pasty, unpalatable, dusty material that can be inhaled and irritate the eyes [17]. It may also induce vomiting; as a result, it affects palatability and, thus, voluntary feed intake. However, it is important to note that fine grinding is essential in human food production.

2.7 Cooking

Cooking of millet is typically completed under steam pressure. The end product can be rolled to produce a product similar to steam-flaked grain. Kate and Singh [22] reported complex changes in physicochemical and functional properties of millets as a result of cooking, including gelatinisation, protein denaturation and the release of bound phenolic and antioxidant compounds. Cooking cereals improves palatability and digestibility slightly compared to unprocessed grains. The duration and temperature of cooking are critical factors to consider when preserving nutritional constituents, particularly minerals and phenolic contents that are heat-labile [22]. McDonald et al. [17] reported improved sorghum grain utilisation through pressure cooking.

2.8 Pelleting

In animal feed production, the feed is ground and then forced through a thick dye (moulder) to form pellets. The method is typically used to reduce the dustiness of feeds. Pelleting results in thorough mixing of the diet as the food is forced through the pelleting machine. In comparison to powdered feeds, pelleting allows for easier mechanical feeding. Pellets can be fed on the ground or in windy conditions since dustiness is reduced. In comparison to meals/mash, pellets do not separate feed ingredients in transit. The advantages of pelleting include that it increases the flow of nitrogen and amino acids into the small intestines [32]. In addition to this, pelleting improves voluntary feed intake, digestibility and feed value. The method is commonly used in rabbit and poultry feeds, among others.

2.9 Malting

Another processing technique commonly used in millets is malting. This is a three-step process carried out in sequence [22]:

i. Steeping (e.g., soaking of grains in water),

ii. Germination (sprouts development and enhancement of enzymatic activity) and

iii. Kilning (e.g., grain drying to stop enzymatic activity). This process dries the grain down to 3–5% moisture and arrests germination. It can be done using natural sun heat. However, where huge volumes of grain are involved, large volumes of hot air, 80–220°C, are blown through the grain bed [33], driving away moisture until desired moisture levels are reached.

Malting enhances the sensory properties, nutritional quality and digestibility of grains while lowering anti-nutrient levels. The kilning process yields malts of varying colours and flavours [33]. Polycyclic aromatic hydrocarbons (PAHs) are known to appear at higher temperatures and are generally considered undesirable in food and beverages because they pose a serious health risk to humans [33]. They are caused by pyrolytic processes that involve the incomplete combustion of wood, organic matter, coal or oil. PAH molecules have varying degrees of carcinogenicity, but all contribute to the overall carcinogenicity of foods exposed to smoking or other sources of PAHs.

2.10 Extrusion

Extrusion is a process of gelatinising and cooking a product until it is fully cooked, resulting in the production of various types of food. This process involves the use of heat or steam and pressure. Grains will move under pressure through an extruder, and by the time it exits (extrusion), the anti-nutritional factors will have been deactivated due to high temperatures in the extruder [16]. This affects protein content because it reacts and changes its levels and structure during the process [34]. Extrusion is a processing technique that reduces the protein content, solubility and water retention of millet varieties. The protein content may decrease due to denaturation of proteins, which has been observed to occur at temperatures above 50°C [17, 29]. Akharume, Santra and Adedeji [34] found that the average denaturation temperature of proteins in Proso millet was 82.1 ± 3.5°C. Hydrophobic properties decrease in response to extrusive processes, but emulsion stability remains unaffected. Emulsion stability influences the colour, texture and succulence of millets [35]. The stability is desirable because it prevents the separation of the oil and water components, resulting in a blending effect that is desirable for maintaining the quality, freshness and integrity of cooked products [36]. Extrusion improves feed value by inactivating anti-nutritional factors, and it may also promote undegraded digestible dietary protein for ruminant animals due to the high-temperature heating involved [24].

3. Effects of processing on nutrients

Millets are usually consumed after being subjected to a variety of processing techniques, such as heating, soaking, dehulling, steaming, roasting, germination, fermentation and other processing methods, which may alter nutritive content. Millets are bestowed with a wealth of nutrients, fibre and antioxidant properties that support and boost metabolism, augment heart health and control blood pressure. Among the nutrients is magnesium, which plays a role in maintaining normal blood pressure. Its adequate intake is associated with a lower risk of hypertension and cardiovascular complications. Millets contain amino acids, lecithin and methionine, which help in bringing down cholesterol levels by eliminating excess fat. The amino acid, tryptophan, in millets lowers appetite and thus assists in weight control [37]. Reduction in weight and cholesterol levels is essential in controlling blood pressure and keeping the heart healthy. These crops help balance cholesterol in the human body [21]. Millets contain essential fats, just the right amount to give our body adequate fat and help prevent excess weight. Millets are low in calories and have a low glycaemic index, making them a good choice for those on a weight-loss diet. The amino acid, threonine, in millets hinders *fat* formation in the liver. The soluble fibre in millets results in highly viscous intestinal contents that possess gelling properties which delay the absorption of carbohydrates by the intestines and thus reduce weight gain [38]. Their dietary fibre keeps you feeling full after eating, and proteins and other nutrients make them a better option to choose for the weight loss journey.

3.1 Effects of processing on protein

There are some changes in millet's nutritional properties with respect to processing methods. Processed millet has, on average, increased protein content of approximately 13.75 g/100 g compared to raw seeds (10.60 g/100 g) [24]. This is due to the concentration of the protein content through the removal of hulls and other processes. Heat-moisture treatment of millet usually has a minimal effect on the protein content. The hydrophobicity of the protein is not altered much by heat-moisture treatments; therefore, there are not many changes to the protein in millet from such processing techniques [39].

During the extrusion process, protein solubility and structure are decreased and disrupted when applied under high pressure and temperature [40]. Germination has been reported to improve the protein content with better water solubility and oil absorption capacity.

3.2 Carbohydrates

In general, processing can destroy starch granules in millet to varying degrees and reduce them as well. With the decrease in particle size of ultra-fine powder, the solubility increases but the gelatinisation temperature decreases; the freeze-thaw stability, enzymolysis and sedimentation properties are also significantly improved [41]. In addition, fermentation with the bacterial strain, *Saccharomyces cerevisiae* reduces starch content in millet varieties with subsequent increase in carbon dioxide and ethanol production throughout the fermentation period. Moreover, pH is significantly reduced, which activates the phytase enzymes, which hydrolysis phytate, thereby reducing the phytate content [41]. This reduction is important as it enhances the bioaccessibility of major nutrients.

During processing, the starch molecules in millet are heated in an aqueous or moist environment, causing them to swell, rupture and burst. The starch gets gelatinised, permitting greater enzymatic digestion by amylases. Cooking increases the digestibility of carbohydrates [42]. Millets are low-glycaemic index (GI) foods and can help keep blood sugar from spiking after meals. They contain indigestible carbohydrates that help control blood sugar, fibre and non-starchy polysaccharides; hence, millets are a good whole grain, especially if you have type 2 diabetes [43]. Type 2 diabetes is when the body makes less insulin than is needed, but in type 1, the body does not make any insulin.

Hydrothermal processing in millet fractionates the starch into amylose and amylopectin fractions, causing a decrease in cold, hot water soluble and hemicellulose B fractions and an increase in pectic polysaccharides, hemicellulose A and cellulosic fractions [44]. Natural fermentation of millets can improve their lower cooking quality, taste, low bioavailability and palatability. Fermentation helps break down nutrients in food, making them easier to digest than their unfermented counterparts. As a result, fermented millets provide many health benefits such as antioxidant, antimicrobial, anti-fungal, anti-inflammatory, anti-diabetic and anti-atherosclerotic activity [45].

Fermentation and germination also affect the amylose content and the structure of amylopectin in millets. These treatments yield more pores in the granule surface and change the crystallinity of the flour. Changes in these structural properties will affect the gelatinisation temperature and application of the treated flour. Igbetar [46] states that soluble sugar is low in millet flour (2.61%), and it increases twofold and sixfold after cooking and fermentation for 72 h respectively. This suggests that fermentation and cooking increase the availability of soluble sugars. There is reduction of total levels of sugar during roasting which is attributed to caramelisation and other reactions [47].

3.3 Fats

Millets are highly nutritive, but their fat content is generally low, which has a positive impact on human health. Processing of the millets results in alterations in the fat content of the millets [48]. For example, there is a reduction in the fat content of millet milk (0.74–0.6%) when compared to that of unprocessed millet milk, which is around (9.1%). The total fat content remains unchanged due to parboiling, which is a heat treatment process. However, some portion of the endosperm fat migrates towards the periphery of the grain, and the oil globules in the aleurone layer are disrupted. During processing, the specified ratios of compositing lead to an increase in the concentration of fat [49]. Pearl millet has omega-3 fatty acids, alpha-linolenic acid, eicosapentaenoic and docosahexaenoic as their fat nutrient [50].

Steeping and fermentation of millets leads to an increase in fat fraction in the treated flour coupled with better water absorption capacity and hygroscopicity. Dey et al. [51] postulate that the effect of processing on some millet varieties is to remove toxins, reduce fat and anti-inflammatory factors due to the presence of millets containing adverse compounds (e.g., hexadecenoic acid, methyl ester, stigmasterol, C-sitosterol and pregnenolone) which exhibit a range of activities (e.g., antimicrobial, nematocidal and anticancer). Other processing techniques such as soaking, grinding or milling have less impact on the fat content present in the millet varieties. The impact varies from one technique to the other and also depends on the nature of the nutrient being tampered with. Finger millet is a good source of essential fatty acids and antioxidants [16]. The fat component in millet will determine the presence of fat-soluble vitamins, and any processing technique that impacts negatively on the fat fraction will inevitably impact on the vitamins also, particularly the fat-soluble ones.

The malting process in millets significantly lowers the fat contents of the malted samples. Raw millet would contain significantly higher fat content that has not undergone any processing. Malting generally improves the digestibility of these fats in millet, and as a result, malted grains have extensively been used in weaning and geriatric foods [16].

3.4 Minerals

Mineral-wise, millet is a rich source of iron, zinc, copper, manganese and potassium, and its mature kernels are rich in vitamin A but deficient in vitamins B and C [52, 53]. The grains consist of iron and calcium in high concentration when compared to other cereal grains [52]. Finger millet grains also contain high amounts of magnesium and phosphorus. Absorption and utilisation of these nutrients in the human body contribute to the reduction of chronic diseases such as lowering of high blood pressure, ischaemic strokes, cardiovascular diseases, cancers, obesity and type II diabetes [49, 54].

Recent studies have shown that some processing methods such as malting, fermentation, decortication, soaking and steaming can improve the bioavailability of these nutrients [37, 46, 47]. Fermentation and germination can improve the bioavailability of iron significantly by around three times and contribute to the requirement of women by around two times. The processing methods, which include grinding, shelling, high pressure, ultra-sound and microwaving, affect the eating quality characteristics and physicochemical properties of millets and cause changes in polyphenol content. Flour from whole grain finger millet was found to have a higher calcium content (325 mg/100 g) than those made from decorticated grain (222 mg/100 g) [48].

The presence of zinc in millet grain was found to be useful in assisting men's health and the immune system. Raw grain was found to contain less zinc than processed grain. Soaking, boiling and germinating pearl millet can reduce phytate and phosphorus content, while also increasing the concentration of calcium, magnesium, iron and zinc [55–57]. These techniques, in particular, impact differently on the contents of different minerals; for instance, a processing technique can have the effect of increasing the bioavailability of one mineral while decreasing or inhibiting that of the other [26].

3.5 Vitamins

Millets contain fat-soluble vitamins, which play an integral role in many physiological processes such as vision, bone health, immune function and coagulation [58]. Millet is rich in B vitamins, especially niacin, pyridoxine and folic acid [59]. These B vitamins are especially important to pregnant and breastfeeding individuals as they aid in foetal brain development and reduce the risk of birth defects. Niacin is essential for the body to convert carbohydrates, fat and alcohol into energy, while pyridoxine is important for normal brain development and a healthy nervous and immune system [59]. Folic acid is essential in the formation of red blood cells and protein synthesis. Soaking helps to increase the nutritional contents of the grain and reduces the cooking time of these millets.

Grinding millet into flour retains its inherent nutritional benefits, providing a good source of vitamins, fibre and essential nutrients. The slow gentle grinding action minimises heat generation, which can help retain the nutritional integrity of the millet flour. As a result, stone ground millet flour may contain higher levels of essential nutrients, such as fibre, vitamins and other minerals, than flour produced using mechanical milling methods [60].

Heat treatments will destroy the heat-labile vitamins in millet, and these include roasting, boiling, parboiling, etc. Cooking and extrusion can reduce the vitamin content; as such, heat treatments should be done over a short time period and under controlled temperatures so that the best outcome is achieved [26].

4. Effects of processing on anti-nutritional factors

Millets contain readily available nutrients and energy sources; hence, scientists, agriculture industries and food security policies are giving more attention to millet processing for its better utilisation [8]. It gets considerable attention because of its nutritional quality related to its high content of dietary fibre, protein and starch patterns, as well as high mineral levels [61]. Millets also contain a considerable amount of phytochemicals that are important in reducing chronic diseases such as cardiovascular diseases, cancer and diabetes [61]. Bioavailability of nutrients is restricted due to the presence of anti-nutritional factors in millets, which include phytic acid, tannins, goitrogens, oxalic acid and trypsin inhibitors. These compounds interfere with mineral bioavailability, carbohydrates and protein digestibility through inhibition of proteolytic and amylolytic enzymes [51]. Anti-nutritional factors in millets interfere with the digestibility of proteins and starch along with reduction in bioavailability of minerals as a result of phenol–protein interactions and metal chelation.

4.1 Phytic acid

Phytic acid is present in the bran portion of cereals as a crystalline globoid. It is the organic form of phosphorous (myoinositol 1,2,3,4,5,6,-hexakis dihydrogen phosphate) occurring in plant constituents as the major portion of phosphorous. They are negatively charged and have a compound that attracts and binds positively charged substances like zinc, iron and calcium minerals, forming insoluble complexes that are unavailable for digestion and absorption [51]. Phytate is an anti-nutrient that impairs the bioavailability of some minerals such as copper, iron, zinc and calcium [62]. Phytic acid has a strong chelating ability and readily forms mono and multivalent cations of potassium, calcium, iron, zinc magnesium and other cations, reducing their bioavailability [63]. Chelates formed with the di- and trivalent metallic particles of minerals are insoluble compounds and are not absorbed in the gastrointestinal tract (GIT) [51]. Krishnan and Meera [26] reported that finger millet grains contain polyphenols and phytates which are known to influence the availability of minerals. Processing reduces the phytic acid concentration in millets, hence reducing the amount of positively charged compounds such as iron, zinc and calcium bound to it and making them bioavailable.

4.2 Phenolic compounds and tannins

Phenolic compounds are a class of secondary metabolites found in plants and are further divided into phenolic acids and polyphenols. Millets contain nine identified phenolic acids: gallic acid, vanillic acid, syringic acid and trans-cinnamic acid [64]. Phenolic compounds interfere with numerous enzymatic systems in humans, especially humans, especially those that control thyroid hormone synthesis [63]. Epidemiological evidence suggests that a diet based on millet, as a staple food, plays a role in the formation of goitre. In South Sudan, goitre prevalence in rural areas was high where 74% of diet is derived from millets causing iodine deficiency [63]. Tannins prevent protein from being digested and phenolic compounds reduce the digestibility of protein and carbohydrates as well as the bioavailability of vitamins. Hydrolysable tannins are susceptible to digestive hydrolysis, which can result in toxic substances [51]. Soaking dissolves phytates and hydrolysable tannins, thereby reducing their concentration and positively impacting feed value [24].

4.3 Effect of processing on anti-nutritional factors in millets

Millets have a high concentration of anti-nutritional factors, which reduce the bioavailability of nutrients. Processing of millet is important to reduce amounts of phytic acid and polyphenols and get the full nutritional benefits of the grain. Processing of the millets has shown promising results in their successful utilisation [65]. Dehulling reduces polyphenols and phytic acid in pearl millet. *In vitro* protein digestibility is increased due to the removal of anti-nutrients, which precipitates proteins when millets undergo decortication. Abrasive decorticated pearl millet showed decreased anti-nutritional compounds (fibre and iron binding phenolics compound); however, high phytate content after decortication might be associated with their occurrence in germ and endosperm [65]. Sharma et al. [66] reported that after soaking millet, protease and amylase activity increased significantly with the increased duration of sprouting. Protein solubility increased with soaking and sprouting, modifying the proximate composition of millet grain by enhancing the hydrolysis of complex insoluble organic compounds present in seeds. Sheethal et al. [67] state that the molar ratio of phytic acid to zinc and phytic acid to iron decreased after the application of different processing methods, especially those associated with fermentation.

5. Effects of processing on bioavailability and bioaccessibility of nutrients

Millets have been reported to be highly nutritious, having high amounts of polyunsaturated fatty acids and high dietary fibre, unavailable carbohydrates and a high satiety effect. They are a very good source of calcium, iron, phosphorus, zinc and potassium [68]. Millets have antimicrobial, anti-diabetic and anti-mutagenic properties due to the presence of polyphenols [69]. This makes them an ideal solution to health problems such as obesity, diabetes and hypertension. Millets, however, have

Processing method	Millet type	Effect	Source
Decortication/ Dehulling	Pearl millet and finger millet	Decreased in dietary fibre, minerals and antioxidant activity Removal of phytate localised in the outer layer increases mineral bioavailability	[73]
Milling	Pearl millet and finger millet	Reduction in anti-nutrients and an enhancement in the bioaccessibility of minerals Increase in protein and starch digestibility	[72]
Hydrothermal/ germination	Pearl millet and finger millet; Sorghum	Enhancement of mineral availability by breaking down anti-nutrients Enhances nutrient content and energy density while reducing phytic acid in food products. Increases *in vitro* bioaccessibility of minerals like calcium, zinc and iron	[71, 74]
Fermentation with lactic acid bacteria	Pearl millet, Finger millet, foxtail millet bran	improves protein digestibility and nutritional value and limits the action of the anti-nutrients, tannins, phenols, phytate and trypsin inhibitors	[71, 72, 75]
Soaking and germination	Finger millet, pearl millet and sorghum	Soaking grains before dehulling and milling reduces phytates tannin and phenolic content. Phytate content is reduced as soaking activates phytase. This in turn increases the availability of Ca, P, Zn, Fe, Cu, K and Mg	[64–66, 76]
Blanching	Pearl millet	Reducing inhibitory the factors phytic acid and polyphenol increases the iron and zinc bioaccessibility	[26, 77]
Malting (germination, milling and sieving)	Pearl millet	Improves the nutritional quality by decreasing tannin and phytic acid levels, making Na, Ca, P, Zn, Fe and Cu more bioaccessible	[78]
Roasting	Pearl millet	Roasting reduces polyphenols in finger millet Increased Na, Ca, P, Zn, Fe and Cu availability when roasted at 70°C, above which Fe and Mn availability decreases	[24, 79]

Table 1.
Effect of processing methods on the nutrient content and the bioaccessibility and bioavailability of nutrients.

tannins, phytates, polyphenols and trypsin inhibitors. These anti-nutrients influence the bioaccessibility and bioavailability of both micro and macronutrients. Shobana et al. [70] reported poor iron availability in some varieties of finger millet due to thigh tannin content.

The absorption of nutrients in the gastrointestinal tract is highly dependent on the release of nutrients from food. The released nutrients are then accessible for absorption in the small intestine. Bioaccessibility refers to the amount of nutrients available for intestinal absorption. The quantity of ingested nutrients, which is then absorbed and utilised is defined as bioavailability [71, 72], and it indicates nutritional effectiveness. Bioavailability is affected by the form in which the nutrient is incorporated, its chemical bonding and its interference with other nutrients, which can either enhance or inhibit absorption or post-absorption metabolism. The nutritional and health benefits of millets are, therefore, subject to bioaccessibility and bioavailability of the nutrients in millets.

The bioaccessibility and bioavailability of nutrients in millets is reliant on the mechanical breakdown of foods and the enzymatic hydrolysis of nutrients. Bioavailability is also influenced by the localisation of nutrients within the food matrix, physical and chemical breakdown of the food and the availability of the lipid phase. Millet processing affects the bioaccessibility and bioavailability of nutrients and the processing methods can result in the transformation of the food matrix and nutrient forms. Decortications, milling, germination, fermentation, malting and roasting have all been reported to improve the organoleptic and nutritional properties of millets. **Table 1** summarises the effects of some processing methods on the bioaccessibility and bioavailability of millets.

6. Value-added millets and their diverse use for food and feed products

Millets are being considered as a valuable and promising solution for the achievement of food security, in the face of population increases and adverse climatic conditions, especially in the developing world. Increased adoption of millets as a food source for all and not just for the poor rural communities will only be possible through the value addition of millet products, in the form of varying dishes such as pasta, baked, flaked and popped products as well as instant food recipes [80, 81]. Millets can totally replace wheat, maize and rice or be blended together with these cereals in the production of varying cakes, pasta, macaroni, vermicelli, noodles, spaghetti and flakes [2, 5, 19]. The processing of millets can result in a number of various products (**Table 2**) for human consumption. Millets are commonly used to make stiff or thin porridge or can be cooked like rice including soup or weaning food [82].

Millets have also been used as livestock feed, including ruminants, pigs and poultry. They have been adopted in the production of animal feed as a source of energy and protein in place of corn. Ground or whole pearl millet can be mixed with maize up to 50% inclusion level. Hassan et al. [16] found that the inclusion of 40% pearl millet in livestock feed increased milk production with no adverse effect on milk quality in dairy cows. The inclusion of pearl millets in animal feed resulted in improved egg quality with higher omega-3 fatty acids and lower omega-6 [84].

A study looked into the effects on broiler chicken performance of using millet hulls in place of wheat offal (0–100%) [85]. The daily feed, weight gain, feed conversion ratio, feed efficiency ratio, gut contents and prime cuts were all found to be similar. Feed costs can be decreased by using millet hulls up to 100% of the

Value-added product	Processing technique
Porridge	Cooking [82]
Puffs	These are made from different puffing machines, giving expanded and crispy products [82]
Popped and flaked products	Flakes are flat round shaped products which are ready to eat and are made using an extruder and a roller flaker machine [81, 82]
Millet cookies and cakes	Baking has been through oven baking of different mixed flour of some of the following pearl millet, finger millet flour combined with usual baking ingredients such as sugar, milk, eggs, vanilla, fat and salt [82, 83]
Millet pasta	Sorghum, finger millet and pearl millet are mixed with a small portion of flour and extruded in a pasta-making machine [81, 82]

Table 2.
Various products from millets for human consumption.

time in broiler chicken diets. The main phenolic components in millet husk are 1-O-p-coumaroylglycerol, apigenin-C-pentosyl-C-hexoside and 1-O-feruloyl-3-O-p-coumaroylglycerol being [86]. Therefore, millet husks, in addition to their acetylcholinesterase and α-glucosidase inhibitory activities, are a valuable source of naturally derived antioxidants [85, 86].

7. Conclusion

Millets are considered to have high nutritional components and health benefits as they are sources of quality protein, dietary fibre, energy and essential minerals. The health benefits from millets are a result of the presence of antioxidant properties exhibited due to the presence of polyphenols such as tannins and some phenolic compounds that can protect the body from oxidative stress. These also act as anti-inflammatory and antiviral agents, thus boosting the immune system. Despite millets being known globally as a source of quality grains, they continue to lack established and improved processing techniques, hence hampering the extent of millet processing and utilisation. However, the potential for millet as a source of valuable health food and feed remains high. Prior to the inclusion of millets in food and feed formulations, it is necessary to address the issues of their digestibility, micronutrient availability and anti-nutritional factors. Millets' food and feed inclusion is mainly dependent on innovation, value-added products, attractive branding and impressive marketing strategies supported by public awareness and an increased sense of ownership.

Conflict of interest

The authors declare no conflict of interest.

Author details

Rumbidzai Blessing Nhara[1], Charity Pisa[2], Ngavaite Chigede[3], Rachel Gwazani[3], Morleen Muteveri[4], Loreen Murimoga[5] and Faith Matiza Ruzengwe[3]*

1 Faculty of Agriculture and Environmental Systems, Department of Livestock Science, University of Zimbabwe, Zimbabwe

2 Department of Soil and Plant Science, Gary Magadzire School of Agriculture and Engineering, Great Zimbabwe University, Masvingo, Zimbabwe

3 Department of Livestock, Wildlife and Fisheries, Gary Magadzire School of Agriculture and Engineering, Great Zimbabwe University, Masvingo, Zimbabwe

4 Department of Physics, Geography and Environmental Science, School of Natural Sciences, Great Zimbabwe University, Masvingo, Zimbabwe

5 Department of BTech, Masvingo Polytechnic, Masvingo, Zimbabwe

*Address all correspondence to: faithruzengwe@gmail.com

References

[1] Ali A, Bhattacharjee B. Nutrition security, constraints, and agro-diversification strategies of neglected and underutilized crops to fight global hidden hunger. Frontiers in Nutrition. 2023;**10**(June):1144439

[2] FAO, IFAD, UNICEF W, WFP. In brief to the state of food security and nutrition in the world 2023. In: In Brief to the State of Food Security and Nutrition in the World 2023. Rome: FAO, IFAD, UNICEF & WFP; 2023

[3] Sabuz AA, Rana MR, Ahmed T, Molla MM, Islam N, Khan HH, et al. Health-promoting potential of millet: A review. Separations. 2023;**10**(2):80

[4] Chaturvedi P, Govindaraj M, Govindan V, Weckwerth W. Editorial: Sorghum and pearl millet as climate resilient crops for food and nutrition security. Frontiers in Plant Science. 2022;**13**(March):11-14

[5] Kheya SA, Talukder SK, Datta P, Yeasmin S, Rashid MH, Hasan AK, et al. Millets: The future crops for the tropics - status, challenges and future prospects. Heliyon. 2023;**9**(11):e22123

[6] Saini S, Saxena S, Samtiya M, Puniya M, Dhewa T. Potential of underutilized millets as nutri-cereal: An overview. Journal of Food Science and Technology. 2021;**58**(12):4465-4477

[7] Sarita SE. Potential of millets: Nutrients composition and health benefits. Journal of Scientific and Innovative. 2016;**5**(2):46-50

[8] Yousaf L, Hou D, Liaqat H, Shen Q. Millet: A review of its nutritional and functional changes during processing. Food Research International. 2021;**142**(17):110197

[9] Amadou I, Gounga ME, Le GW. Millets: Nutritional composition, some health benefits and processing - a review. Emirates Journal of Food and Agriculture. 2013;**25**(7):501-508

[10] Emkani O, Bonastre RS. Effect of lactic acid fermentation on legume protein properties, a review. Fermentation. 2022;**8**(244):1-43

[11] Jan S, Kumar K, Yadav AN, Ahmed N, Thakur P, Chauhan D, et al. Effect of diverse fermentation treatments on nutritional composition, bioactive components, and anti-nutritional factors of finger millet (*Eleusine coracana* L.). Journal Of Applied Biology & Biotechnology. 2022;**10**(May):46-52

[12] Kumar A, Tripathi MK, Joshi D, Kumar V. Millets and millet technology. In: Millets and Millet Technology. Vol. July. Singapore: Springer; 2021. pp. 1-438

[13] Hassan ZM, Manyelo TG, Selaledi L, Mabelebele M. The effects of tannins in monogastric animals with special reference to alternative feed ingredients. Molecules. 2020;**25**(4680):1-17

[14] Cisse RS, Hamburg JD, Freeman ME, Davis AJ. Using locally produced millet as a feed ingredient for poultry production in sub-Saharan Africa. Journal of Applied Poultry Research [Internet]. 2017;**26**(1):9-22. DOI: 10.3382/japr/pfw042

[15] Baurhoo N, Baurhoo B, Mustafa AF, Zhao X, Mustafa IAF, Zhao X. Comparison of corn-based and Canadian pearl millet-based diets on performance, digestibility, villus morphology, and digestive microbial populations in broiler chickens. Poultry Science. 2011;**90**(3):579-586

[16] Hassan ZM, Sebola NA, Mabelebele M. The nutritional use of millet grain for food and feed: A review. Agriculture & Food Security. 2021;**10**(1):1-14

[17] McDonald P, Edwards RA, Greenhalgh JFD, Morgan CA, Sinclair LA, Wilkinson RG. Animal nutrition. 7th ed. Upper Sadle River, New Jersey, USA: Prentice Hall; 2010

[18] Swami SB, Thakor NJ, Gurav HS. Effect of soaking and malting on finger millet (*Eleusine Coracana*) grain. Agricultural Engineering International: CIGR Journal. 2013;**15**(March):194-200

[19] FAO. Millets Recipe Book. Rome: FAO; 2023. International Year of Millets 2023

[20] Kaukovirta-Norja A, Wilhelmson A, Poutanen K. Germination: A means to improve the functionality of oat. Agricultural and Food Science. 2004;**13**(1-2):100-112

[21] Saleh ASM, Zhang Q, Chen J, Shen Q. Millet grains: Nutritional quality, processing, and potential health benefits. Comprehensive Reviews in Food Science and Food Safety. 2013;**12**(3):281-295

[22] Kate A, Singh A. Processing technology for value addition in millets. In: Millets and Millet Technology. Singapore: Spinger Nature; 2021. pp. 239-254

[23] Hassan S, Hussain MB, Waheed M, Ahmad K, Kassymov S, Shariati MA, et al. Effect of germination processing on bioactive compounds of cereals and legumes. In: Egbuna C, Dable-Tupas G, editors. Functional Foods and Nutraceuticals. Switzerland: Springer Nature; 2020. pp. 283-306

[24] Wang H, Fu Y, Zhao Q, Hou D, Yang X, Bai S, et al. Effect of different processing methods on the millet polyphenols and their anti-diabetic potential. Frontiers in Nutrition. 2022;**9**(780499):1-8

[25] Ikwebe J, Mayel MH, Tatah SV, Mamman E, Sabo DJ. Effects of dehulling on the levels of micronutrients in maize, millet and sorghum grains. Trends in Science & Technology Journal. 2020;**6**(May):220-222

[26] Krishnan R, Meera MS. Pearl millet minerals: Effect of processing on bioaccessibility. Journal of Food Science and Technology. 2018;**55**(9):3362-3372

[27] Lobo V, Patil A, Phatak A, Chandra N. Free radicals, antioxidants and functional foods: Impact on human health. Pharmacognosy Reviews. 2010;**4**(8):118-126

[28] Mannuramath M, Yenagi N. Optimization of hydrothermal treatment for little millet grains (*Panicum miliare*). Journal of Food Science and Technology. 2015;**52**(11):7281-7288

[29] Sugiharto S. The use of sprouted grains as dietary feed ingredients for broilers - a brief overview. Livestock Research for Rural Development [Internet]. 2021;**33**(3):1-5. Available from: http://www.lrrd.org/lrrd33/3/sgh_u3338.html

[30] Sow MS, Sarr F, Mertz C, Cisse M, Fall SM, Diallo D, et al. Effects of fermentation, germination, roasting and mono-screw extrusion cooking on the phytate and iron contents of millet souna produced in Senegal (*Pennisetum glaucum*). Journal of Nutrition Health and Food Science. 2019;**7**(3):1-7

[31] Gombwe C, Chigede N, Hungwe T, Nhara RB, Dahwa E, Muvhuringi PB, et al. Feed Management for Smallholder pig Farming Systems in Zimbabwe.

In: Poshiwa X, Chary RG, editors. Climate Change Adaptations in Dryland Agriculture in Semi-Arid Areas [Internet]. 1st ed. Zimbabwe: Springer Nature; 2022. pp. 301-310. DOI: 10.1007/978-981-16-7861-5_22

[32] Mpofu IDT. In: Manyanhaire IO, Masama E, editors. Introduction to Animal Nutrition. 1st ed. Harare, Zimbabwe: Zimbabwe Open University; 2015

[33] Habschied K, Kartalović B, Kovačević D, Krstanović V, Mastanjević K. Effect of temperature range and kilning time on the occurrence of polycyclic aromatic hydrocarbons in malt. Food. 2023;**12**(3):1-8

[34] Akharume F, Santra D, Adedeji A. Physicochemical and functional properties of proso millet storage protein fractions. Food Hydrocolloids. 2019;**108**:1-28

[35] Jeong Y, Han Y. Effect on the emulsification stability and quality of emulsified sausages added with Wanggasi-Chunnyuncho (*Opuntia humifusa* f. *Jeollaensis*) fruit powders. Food Science of Animal Resources. 2019;**39**(6):953-965

[36] Atchley C. Baking and Snack. 2019. What functionality can emulsifiers bring to baked foods? Available from: https://www.bakingbusiness.com/articles/48188-what-functionality-can-emulsifiers-bring-to-baked-foods#:~:text="Emulsifiershelpmaintainthequality,managementdirector%2Cemulsifiers%2CCorbion [Accessed: February 29, 2024]

[37] Sk M, Sudha K. Functional and phytochemical properties of finger millet (*Eleusine coracana* L.) for health. International Journal of Pharmaceutical, Chemical and Biological Sciences. 2012;**2**(4):431-438

[38] Wolever TMS, Jenkins DJA, Ocana AM, Rao VA, Collier GR. Second-meal effect: Low-glycemic-index foods eaten at dinner improve subsequent breakfast glycemic response. The American Journal of Clinical Nutrition. 1988;**48**(4):1041-1047

[39] Thilagavathi T, Banumathi P, Kanchana S. Effect of heat moisture treatment on functional and phytochemical properties of native and modified millet flours. Plant Archives. 2015;**15**(1):15-21

[40] Siddharth M. Determination of physical characteristics of extruded snack food prepared using little millet (*Panicum sumatranc*) based composite flours. International Journal for Research in Applied Science and Engineering Technology (I JRASET). 2014;**2**(Vii):213-218

[41] Yang T, Ma S, Liu J, Sun B, Wang X. Influences of four processing methods on main nutritional components of foxtail millet: A review. Grain & Oil Science and Technology. 2022;**5**(3):156-165

[42] Bedin AC, Bach D, da Silva Junges MF, Lacerda LG, Demiate IM. Influence of cooking method on the *in vitro* digestibility of starch from sweet potato roots. Brazilian Archives of Biology and Technology. 2023;**66**:1-11

[43] Agrawal P, Singh BR, Gajbe U, Kalambe MA, Bankar M. Managing diabetes mellitus with millets: A new solution. Cureus. 2023;**15**(9):e44908

[44] Dharmaraj U, Malleshi NG. Changes in carbohydrates, proteins and lipids of finger millet after hydrothermal processing. LWT. 2011;**44**(7):1636-1642

[45] Rao D, Sangappa Vishala GD, Arlene Christina VA, Tonapi BA. Technologies of Millet Value Added Products. India:

ICAR-Indian Institute of Millets Research; 2016. 48 p

[46] Igbetar BD. Impact of Processing on the Carbohydrate Quality and Digestibility of Pearl Millet (*Pennisetum glaucum*). United Kingdom: The University of Leeds School of Food Science and Nutrition; 2019

[47] Tumwine G, Atukwase A, Tumuhimbise GA, Tucungwirwe F, Linnemann A. Production of nutrient-enhanced millet-based composite flour using skimmed milk powder and vegetables. Food Science & Nutrition. 2019;**7**(1):22-34

[48] Kindiki MM, Onyango A, Kyalo F. Effects of processing on nutritional and sensory quality of pearl millet flour. Food Science and Quality Management. 2015;**42**:13-20

[49] Ramashia SE, Anyasi TA, Gwata ET, Meddows-Taylor S, Jideani AIO. Processing, nutritional composition and health benefits of finger millet sub-Saharan Africa. Food Science and Technology. 2019;**39**(2):253-266

[50] Nanje Gowda NA, Siliveru K, Vara Prasad PV, Bhatt Y, Netravati BP, Gurikar C. Modern processing of Indian millets: A perspective on changes in nutritional properties. Food. 2022;**11**(4):1-18

[51] Dey S, Saxena A, Kumar Y, Maity T, Tarafdar A. Understanding the antinutritional factors and bioactive compounds of Kodo millet (*Paspalum scrobiculatum*) and little millet (*Panicum sumatrense*). Journal of Food Quality. 2022;**2022**(2016):1578448

[52] Vinoth A, Ravindhran R. Biofortification in millets: A sustainable approach for nutritional security. Frontiers in Plant Science. 2017;**8**(January):1-13

[53] Nambiar VS, Dhaduk JJ, Sareen N, Shahu T, Desai R. Potential functional implications of pearl millet (*Pennisetum glaucum*) in health and disease. Journal of Applied Pharmaceutical Science. 2011;**1**(10):62-67

[54] Kaur KD, Jha A, Sabikhi L, Singh AK. Significance of coarse cereals in health and nutrition: A review. Journal of Food Science and Technology. 2014;**51**(8):1429-1441

[55] Gupta RK, Gangoliya SS, Singh NK. Reduction of phytic acid and enhancement of bioavailable micronutrients in food grains. Journal of Food Science and Technology. 2015;**52**(2):676-684

[56] Dave S, Yadav BK, Tarafdar JC. Phytate phosphorus and mineral changes during soaking, boiling and germination of legumes and pearl millet. Journal of Food Science and Technology. 2008;**45**(4):344-348

[57] Akanbi T, Timilsena Y, Scientific TC, Dhital S. Bioactive Factors and Processing Technology for Cereal Foods. Vol. 2019. Heidelberg, Germany: Springer; 2021

[58] Kaur P, Purewal SS, Sandhu KS, Kaur M, Salar RK. Millets: A cereal grain with potent antioxidants and health benefits. Journal of Food Measurement and Characterization. 2019;**13**(1):793-806

[59] Dayakar Rao B, Bhaskarachary K, Christina GA, Devi GS, Vilas AT. Nutritional and health benefits of millets. Ayushdhara. 2017;**10**(5):859-868

[60] Shahidi F, Chandrasekara A. Millet grain phenolics and their role in disease risk reduction and health promotion: A review. Journal of Functional Foods. 2013;**5**(2):570-581

[61] Xiang J, Apea-Bah FB, Ndolo VU, Katundu MC, Beta T. Profile of phenolic compounds and antioxidant activity of finger millet varieties. Food Chemistry. 2019;**275**(April):361-368

[62] Atuna RA, Ametei PN, Bawa AA, Amagloh FK. Traditional processing methods reduced phytate in cereal flour, improved nutritional, functional and rheological properties. Scientific African. 2022;**15**(December):e01063

[63] Boncompagni E, Orozco-Arroyo G, Cominelli E, Gangashetty PI, Grando S, Tenutse Kwaku ZT, et al. Antinutritional factors in pearl millet grains: Phytate and goitrogens content variability and molecular characterization of genes involved in their pathways. PLoS One. 2018;**13**(6):1-30

[64] Devi PB, Vijayabharathi R, Sathyabama S, Malleshi NG, Priyadarisini VB. Health benefits of finger millet (*Eleusine coracana* L.) polyphenols and dietary fiber: A review. Journal of Food Science and Technology. 2014;**51**(6):1021-1040

[65] Singh N, Kumari P, Gulati J, Bassi N. Nutrient rich Kodo millet, importance and value addition: An overview. The Pharma Innovation Journal. 2023;**12**(6):3713-3719

[66] Sharma B, Gujral HS. Modifying the dough mixing behavior, protein & starch digestibility and antinutritional profile of minor millets by sprouting. International Journal of Biological Macromolecules. 2020;**153**:962-970

[67] Sheethal HV, Baruah C, Subhash K, Ananthan R, Longvah T. Insights of nutritional and anti-nutritional retention in traditionally processed millets. Frontiers in Sustainable Food Systems. 2022;**5**(February):1-14

[68] Dykes L, Rooney LW. Phenolic compounds in cereal grains and their health benefits. Cereal Foods World. 2007;**52**(3):105-111

[69] Viswanath V, Urooj A, Malleshi NG. Evaluation of antioxidant and antimicrobial properties of finger millet polyphenols (*Eleusine coracana*). Food Chemistry. 2009;**114**(1):340-346

[70] Shobana S, Krishnaswamy K, Sudha V, Malleshi NG, Anjana RM, Palaniappan L, et al. Finger millet (Ragi, *Eleusine coracana* L.). A review of its nutritional properties, processing, and plausible health benefits. In: Advances in Food and Nutrition Research. 1st ed. Vol. 69. Amsterdam, Netherlands: Elsevier Inc; 2013. pp. 1-39

[71] Anagha KK. Millets: Nutritional importance, health benefits, and bioavailability: A review. The Pharma Innovation Journal. 2023;**12**(8):223-227

[72] Tharifkhan SA, Perumal AB, Elumalai A, Moses JA, Anandharamakrishnan C. Improvement of nutrient bioavailability in millets: Emphasis on the application of enzymes. Journal of the Science of Food and Agriculture. 2021;**101**(12):4869-4878

[73] Cilla A, Barberá R, López-García G, Blanco-Morales V, Alegría A, Garcia-Llatas G. Impact of processing on mineral bioaccessibility/bioavailability. In: Innovative Thermal and Non-Thermal Processing, Bioaccessibility and Bioavailability of Nutrients and Bioactive Compounds. Cambridge, United Kingdom: Elsevier Inc.; 2019. pp. 209-239

[74] Singh A, Gupta S, Kaur R, Gupta HR. Process optimization for anti-nutrient minimization of millets. Asian Journal of Dairy and Food Research. 2017;**36**(4):322-326

[75] Geervani P, Vimala V, Pradeep KU, Devi MR. Effect of processing on protein digestibility, biological value and net protein utilization of millet and legume based infant mixes and biscuits. Plant Foods for Human Nutrition. 1996;**49**(3):221-227

[76] Chandra B, Viswavidyalaya K. Millets-the super food and it nutritional and health benefits. Research and Reviews: Journal of Environmental Sciences. 2023;**5**(2):26-31

[77] Krishnan R, Dharmaraj U, Malleshi NG. Influence of decortication, popping and malting on bioaccessibility of calcium, iron and zinc in finger millet. LWT. 2012;**48**(2):169-174

[78] Kumar V, Sinha AK, Makkar HPS, Becker K. Dietary roles of phytate and phytase in human nutrition: A review. Food Chemistry. 2010;**120**(4):945-959

[79] Singh Scholar N, David Professor J, Thompkinson Professor Emeritus D, Sagar Seelam Scholar B, Rajput Scholar H, Morya Scholar S, et al. Effect of roasting on functional and phytochemical constituents of finger millet (*Eleusine coracana* L.). The Pharma Innovation Journal. 2018;7(4):414-418

[80] Mote M. Value added products of millet in India and it's branding. Just Agriculture. 2023;**3**(5):3-6

[81] Saraswatt HR. Value addition influenced in millet products - a review. Austin Journal of Nutrition and Food sciences. 2022;**10**(1):1161

[82] Dharmaraj U. Value Added Products from Millets, Mysore Major Cereals, Coarse Cereals and Millets. Mysore: CSIR-CFTRI; 2011

[83] Kumar V, Shekhawat P, Banga P. Millets and their value-added products. Just Agriculture. 2023;**3**(5):3-6

[84] Hassan ZM. The nutritional use of millet grain for food and feed. Agriculture & Food Security. 2021;**10**:16

[85] Maidala A, Bakoji S. Nutritional evaluation of millet hulls as a source of fibre in the diets of broiler chickens. IIARD International Journal of Biological & Medical Research. 2016;**2**(1):1-9

[86] Wang C, Li Z, Xiang J, Johnson JB, Zheng B, Luo L, et al. From foxtail millet husk (waste) to bioactive phenolic extracts using deep eutectic solvent extraction and evaluation of antioxidant, acetylcholinesterase, and α-glucosidase inhibitory activities. Food. 2023;**12**(6):1144

Chapter 3

Cereal By-Products Valorization in Bakery, Pastry, and Gastronomy Products Manufacturing

Maria Simona Chiș and Anca Corina Fărcaș

Abstract

Cereals represent one the most famous crops worldwide, covering more than 20% of the human daily diet. Through their processing, several agro-food chain by-products are generated, emphasizing an urgent need for further valorization considering economic, social, and environmental factors. The ever-increasing demand for food requires new, healthy, and sustainable products. Therefore, the present chapter aims to highlight the main possibilities for cereal by-products valorization in bakery, pastry, and gastronomy products. Fermentation of the cereal by-products with lactic acid bacteria, optimization of the added by-product percentages in new functional products, extrusion process, and food fortification will be the main topics of the proposed chapter. The influence of the cereal by-products addition on human health will be also discussed.

Keywords: cereals, by-products, fermentation, lactic acid bacteria, extrusion, anti-nutrients, bioactive compounds

1. Introduction

Cereals are the main worldwide crop with a planting area of more than 73% of the total global harvested area [1]. The term 'cereals' includes nine *Gramineae* family members, species such as wheat (*Triticum*), barley (*Hordeum*), rye (*Secale*), rice (*Oryza*), oat (*Avena*), millet (*Pennisetum*), sorghum (*Sorghum*), corn (*Zea*), and triticale [2]. Cereals are considered staple crops and are primarily cultivated for their edible grain or kernels [3]. More precisely, the cereal fruit is a caryopsis composed of a pericarp (the fruit coat) and a seed. The fruit coat tightly adheres to the seed coat, encasing the germ and endosperm within the remainder of the seed. Adjacent to the pericarp is the aleurone layer, characterized by its richness in protein and minerals. The endosperm, constituting the substantial central part of the kernel, is primarily composed of starch, while the germ or embryo is the smaller structure located at the lower end of the kernel (**Figure 1**) [3].

Wheat, rice, and maize are the main contributors to the 12% increasement of the global food production till 2025 as indicated by Food and Agriculture Organization (FAO) statistics, [4]. In line with this, Fărcaș et al. [5] stated that more than 90% of

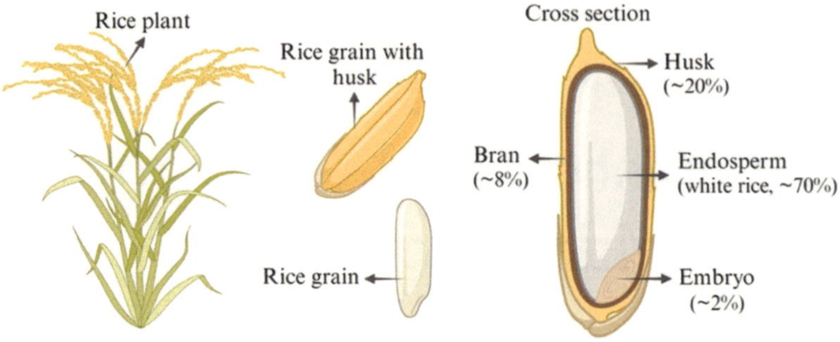

Figure 1.
Rice plant and cross section of the rice grain.

cereal consumption is represented by rice, wheat, barley, and maize. From 2008 to 2016, the annual worldwide production of cereals averaged around 3200 million tons, encompassing grains such as wheat, maize, sorghum, barley, rice, oat, rye, and millet [6]. During 2020–2021, the worldwide cereal production was formed mainly of corn, with a percentage of 42%, wheat 29%, and rice and barley with values of 19% and 6%, respectively [7].

Cereal grains have been a fundamental element of the human diet for millennia. Their processing is a crucial aspect of the food production chain, involving a complex process such as dry milling (for wheat and rye), pearling (for rice, oat, and barley), wet milling (for corn and wheat), and malting (for barley, corn, and wheat), [6]. In the process of dry milling, the outer fibrous materials and germ are separated, resulting in by-products of the grain endosperm, whilst pearling is a dry milling abrasive technique that systematically eliminates the seed coat (testa and pericarp), aleurone and subaleurone layers, as well as the germ, to achieve polished grain (as seen in rice, oat, and barley), along with by-products containing a high concentration of bioactive compounds [2].

Conversely, wet milling is primarily employed in the manufacturing of starch and gluten, generating steep solids (abundant in nutrients valuable for the pharmaceutical industry), germ (utilized in the oil-crushing industry), and bran as by-products. Malting, designed for the production of beer and other alcoholic beverages, involves the consumption of fermentable sugars and starch in the grain (typically barley) by enzymes, resulting in a by-product entitled spent grain [2].

Throughout cereal processing, by-products with distinct physical states and chemical compositions are simultaneously generated [6]. Globally, approximately 12.9% of total food waste is produced through cereal processing [5], Europe and North America claiming a total cereal food chain loss production between 10 and 12%, meantime, Asia recorded a percentage of up to 18% [8]. Only in Europe, the total amount of cereal waste is approximately 40,000–45,000 tons/year [8].

The cereal milling has as the main objective the flour manufacturing and generates various by-products such as bran, hull, germ, husk, fiber, protein, and broken grains [6]. The cereals bran and germ fractions, also coined the "dark matter of nutrition" are abundant in vitamins, minerals, dietary fiber, and thousands of phytochemicals [9].

Bran is the main by-product generated from the cereal milling process and typically refers to the outer layers of the grain. Its composition is highly variable and influenced by factors such as grain type, kernel size, shape, maturity, germ size, pericarp thickness, grain storage duration and condition, pre-milling grain conditioning, milling process, and the machinery employed during milling [3]. The percentage of the bran obtained after wheat milling is about 15%, whilst, the barley yield of milling by-products is estimated to be around 30–40% [3]. With respect to wheat, given that over 650 million tons are produced and processed annually, the associated quantity of wheat bran was approximated to be 150 million tons per year [10].

On the other side, the processing of rough rice, undertaken to produce milled edible rice commonly known as white rice, results in by-products, namely rice hull (constituting 20% of rough rice) and rice bran and germ (making up 10% of rough rice), [8].

The dry milling maize process led to the production of maize bran fraction and further to maize germ, a by-product used in the oil production, whilst, wet maize milling generates by-products such as maize fiber and maize protein [11]. Maize germ represents 10–12% of the whole corn kernel, meantime, corn bran could be varied from 5 to 6% [12]. Nowadays, there is an increasing interest for the bakery companies to use maize flour in product manufacturing, therefore, the maize bran demand is highly increasing [11].

Apart from the dry milling maize process, the barley malting process led to spent malt rootlets by-product generation, meantime, the beer manufacturing process produced a huge amount of brewer spent grain, spent hops, and surplus yeast [7].

Therefore, considering that the main worldwide crops are corn, wheat, rice, and barley, the chapter will characterize the chemical composition of the main by-products generated after selected cereals processing. Afterward, it will emphasize modern strategies to increase the bioaccessibility of bioactive compounds, show their further use in the manufacturing of bakery, pastry, and gastronomy products, and review the main positive influences on human health.

2. General composition and bioactive compounds in cereal by-products

Wheat bran is a rich source mainly in carbohydrates, with an amount more than 57% (from which starch, hemicelluloses, and cellulose are the main representants), protein in a range between 13 and 18%, lipids: 3–4%, 3–8% ash, [10], dietary fiber along with a diverse array of biologically active compounds, including alkylresorcinol, ferulic acid, β-glucan (ranging from 22 to 27 g/100 g dry weight), and arabinoxylan (ranging from 22.4 to 29.80 g/100 g dry weight) [7]. Wheat bran and oat bran contain a high amount of phenolic acids such as 4527 and 4190 μg/g, respectively [1]. Wheat germ has a lipid composition in a range of 10–15 g/100 g dry weight, protein amount between 26 and 35 mg/100 g dry weight, fiber with values between 15 and 45 g/100 g dry weight as well as bioactive compounds such as thiamin, riboflavin, tocopherols, but also phytosterols and carotenoids [7].

With respect to corn bran's chemical composition, it contains 10–13% proteins, lipids (2–3%), starch (9–23%), 2% ash, and a high amount of dietary fiber (76–90%), [8, 13]. Typically, corn bran contains various phenolic acids, with ferulic acid being the primary one, but also vanillic, caffeic, p-coumaric, and p-hydroxybenzoic acids have been identified in significant amounts [8].

On the other hand, rice bran (composed of bran layer and rice germ) is a good source of fatty acids, minerals, dietary fibers, protein, and several phytochemical compounds, claimed to play a significant role in biological activities such as antioxidants or anti-inflammation [14]. It is mainly composed of carbohydrates in a range between 30 and 50%, lipids in a higher amount compared with wheat bran (up to 19%), and protein around 14% [10]. The results are in line with Sapwarobol et al., [15] who mentioned a 50% carbohydrate level, fat in an amount of 20%, protein 15%, and dietary fiber. The bran derived from brown rice is recognized for its prebiotic potential, it contains beneficial compounds such as dietary fibers, essential fatty acids, polyphenols, and antioxidants [11]. Rice bran is considered to have antioxidative, hypocholesterolemic, and hypoallergenic properties, being an important source of nutrients. On the other side, black rice contains a better amount of minerals, higher amino acids amount, and phytochemicals compared to the white rice. Black rice germ and bran contain flavonoids, vitamin E, and γ- oryzanol with positive benefits such as anti-inflammatory and antioxidants characteristics [14].

Additionally, it is important to mention the rich chemical composition of the main by-products from the beer manufacturing and barley malting processes such as brewer spent grain and spent malt rootlets, respectively. Brewer spent grain is claimed by the literature as having high fiber amount (up to 70%), protein in a range between 25 and 30%, and lipid up to 10% [10]. It has a big advantage of being available throughout the whole year, at a low or no cost, and being produced in large amounts by industrial or small breweries [16]. Brewer spent grain is a rich source of phenolic compounds, mainly caffeic, *p*-coumaric, protocatechuic and ferulic acids, catechin, and other derivates [17]. It exhibited a high amount of protein and amino acids such as threonine, cysteine, proline, glutamine, and leucine [17]. It is an extremely rich source of fiber (up to 70%) mainly formed of cellulose, lignin, hemicellulose, and vitamins such as choline, niacin, pantothenic acid, and riboflavin [18]. It is worthy to mention that minerals such as Ca (calcium), Na (sodium), K (potassium), Mg (magnesium), Fe (iron), and Ni (nickel) are important compounds present in brewer spent grain [19]. Soil fertility, production, barley variety, harvesting time, mashing, and malting conditions are the main factors that could influence the brewer spent grain composition [20].

Spent malt rootlets are a rich source of protein (35 g/100 g), a source of minerals (total ash 5.98 g/100 g), total fiber (36.64 g/100 g) and soluble fiber (1.24 g/100 g), [21], and represents the main by-products generated during the barley malting process. Spent malt rootlets represent 5% of the total malt weight [21] and it is a rich source of essential and non-essential amino acids, phenolic compounds, mono and di-saccharides sugars, and insoluble dietary fiber [22]. Spent malt rootlets are rich in ferulic and p-coumaric acids, being a natural source of antioxidants [23]. Moreover, it is a rich source of macrominerals such as Ca (0.19%), P (phosphorus) (0.69%), Mg (0.17%), and K (1.17%) and microminerals such as zinc (Zn), manganese (Mn), and copper (Cu) with values of 64 ppm, 36 ppm, and 9 ppm, respectively [24]. It is worth also mentioning its rich composition in vitamin E, B group vitamins, and peptides [25].

3. Advancements in nutrient bioaccessibility of cereal by-products

This part of the chapter will go over the main methods used for increasing bioaccessibility and bioavailability of cereal bioactive compounds, as presented in **Figure 2**. Fermentation processes with lactic acid bacteria (the submerged and the solid-state), enzymatic treatments, and extrusion will be the main discussed topics. The cereal processing industry is actively seeking alternative methods to reduce the generation of by-products and waste.

Consequently, several efforts are being made to explore innovative solutions and applications for the valorization of these by-products. This involves their reprocessing to extract valuable compounds or undergo chemical, microbiological, and enzymatic conversions [11] together with their reuse in different food manufacturing processes.

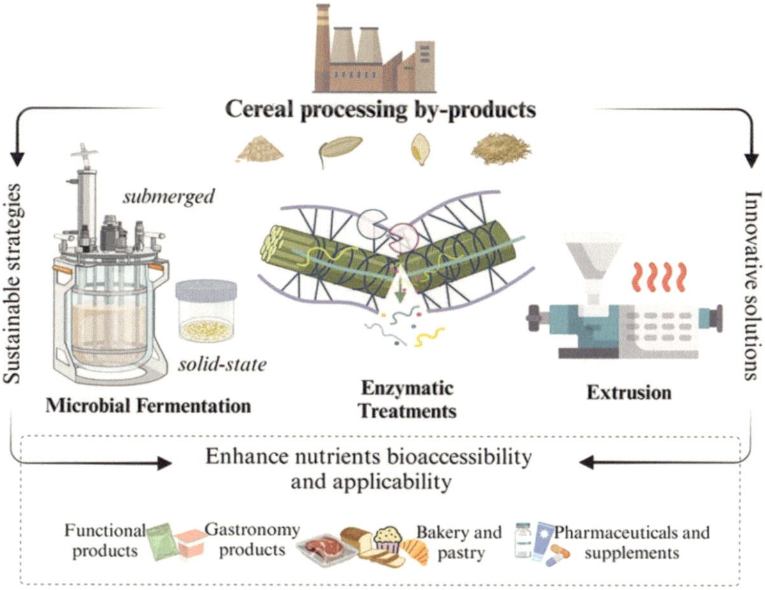

Figure 2.
Pre-treatments used for cereal processing by-products.

Unfortunately, flours with high bran content produce final baked goods with poor loaf volume, dense texture, bitter taste, and dark color [26]. The negative properties of bran wheat on the final baked product are mainly due to the interaction between gluten and other bran components such as antioxidants, phenolics, dietary fiber, and enzymes [27]. On the other hand, it seems that also bran particle size or even water absorption capacity together with the fiber water competition could produce an insufficient hydration of starch and gluten that could negatively influence the wheat bran characteristics [27, 28]. Brewer spent grain is another cereal by-product generated during beer manufacturing that still has a limited exploitation as food raw materials mainly because of its poor technological properties, low sensory performance, and astringent flavor [28].

From the nutritional point of view, cereal bran bioactive compound has a low bioaccesibilty and bioavailability. Bioaccesibility is defined as the release amount of bioactive compounds before their absorption in the small intestine, while bioavailability refers to the amount of bioactive compounds that could be rich into the blood flow, organs, and different tissues [19].

Generally, the low phenolic acids and dietary fibers' low bioaccesibilty are due to their entrapping in the cell wall structures. Moreover, the cereal proteins' bioavailability is limited to several factors such as the presence of antinutritional phytate-protein complexes and the layers' structure mainly formed of insoluble carbohydrates and lignin [29]. Furthermore, compounds such as phytic acid or tannins are present in the cereal outermost tissues being recognized as anti-nutritive compounds [30]. It is already known that phytate forms complexes with minerals present in cereals and seeds, leading to reduced solubility, absorption, and digestibility of minerals. This consequently lowers the bioavailability of these compounds [31]. Moreover, phytic acid content of wheat which could vary in a range between 5.4–8.4% might even reduce the protein nutritional value by absorbing bivalent ions such as Ca^{2+}, Zn^{2+}, Mg^{2+}, or Fe^{2+}, respectively [32].

To overcome this challenge, new milling methods, and new pretreatments such as fermentation and enzymatic ones are highly required to enhance the bran cereal's nutritional potential [29]. Bioprocessing of bran could represent a useful tool and refer to the exploitation of the biological activity exhibited by cells or their components, including microbes and enzymes [33]. In line with this, Hartikainen et al., [33] stated that the use of a big percentage addition of bran wheat in bakery products manufacturing could negatively affect the rheological properties of the dough such as the free expansion, that could further affect the final baked product volume. Moreover, the gluten structure will be affected due to bran action against aggregation of gluten proteins and the bran addition could accelerate the bread staling process [33].

Significant research has investigated the use of lactic acid bacteria to enhance the value of cereal by-products. For instance, rice bran fermented with *Lactobacillus plantarum* at a temperature of 33°C, for 12.5 hours increases the total phenolic content and leads to various aroma volatile compounds [34]. *Lactiplantibacillus plantarum* FST 1.7, *Limosilactobacillus reuteri* R29, *Leuconostoc citreum* TR116, *Lactobacillus amylovorus* FST2.11, and *Weissella cibaria* MG1 were successfully grown on brewer spent grain substrate and fermentation lead to an enrichment of oil-binding capacity, emphasizing fermentation as a promising processing aid for brewer spent grain valorization [35].

Another tool to enrich the chemical composition of cereal by-products is the use of enzymes. Enzymatic treatments offer an alternative approach to mitigate the adverse effects of conventional extraction methods, which typically involve chemical degradation or disruption of cell wall plant matrices through acid or base hydrolysis. This alternative method aims to improve the extractability yield and antioxidant activity of phenolics [11]. Enzymes could increase the bioavailability of ferulic acid. For instance, esterase is able to release ferulic acid from fiber complex in both human and rat intestinal mucosa [31]. In line with this, a sustainable method for brewer spent grain ferulic acid extraction is the use of enzymatic hydrolysis with feruloyl esterase. The enzymatic process has several advantages such as higher extraction yields, environmentally friendly, and process scalability [36]. Ferulic acid recovery from brewer spent grain could be significantly improved through an optimum enzyme dosage, optimization of the incubation time, and parameters such as temperature and pH [36].

Enzymatic hydrolysis is a well-known method used to overcome the insolubility of brewer spent grain protein, with no negative effect on the amino acids chain [37]. Recently, Bazsefidpar et al. [37] showed that the main drawbacks of brewer spent grain such as the low protein solubility or color, could be improved through using different enzyme proteases such as flavoenzyme, alcalase, papain, or pancreatin obtained from different sources such as fungal, bacterial, plants, or even animals.

The bioprocessing of wheat bran through fermentation with *Lactobacillus brevis* E95612 and *Kazachstania exigua* C81116 combined with endoglucanase, xylanase, and β-glucanase hydrolytic enzymes caused a breakdown of the cell structures and increased with more than 11-fold the solubility of arabinoxylans [29]. *Lactiplantibacillus plantarum* and *Leuconostoc pseudomesenteroides* used for the brewer spent grain fermentation increased the phenolic profile of the sample, mainly in vanillin, syringic, and ferulic acid content [28]. *Lactobacillus rhamnosus* ATCC 7469 strain was used in the fermentation of brewer spent grain with an addition of spent malt rootlets in a range between 10 and 50%, aiming to enhance lactic acid production. During the fermentation process it was observed an increasement of the minerals such as iron, zinc, magnesium, or manganese, and the highest lactic acid concentration was reached with an addition of 50% malt rootles extract [25].

On the other side, solid-state fermentation with *Saccharomyces cerevisiae* commercial baker's yeast proved to be a useful technique to increase the cereal bran phenolic extraction yield and improve their antioxidant activity. After 4 days of wheat and oat bran fermentation, the extractable total phenolic content increased with 112 and 83%, respectively [38].

Lactobacillus rhamnosus used for bran solid state fermentation increased threefold the arabinoxylans solubility, decreased with 37% the phytate content, and was able to metabolize the phenolic compounds through breaking the connection with cell-wall polysaccharides [30]. Recently, *Fusarium oxysporum* f. sp. *Lycopersici* was used for the brewer spent grain solid-state fermentation and successfully increased the release of arabinoxylans, free phenolic compounds, phenolic acids, and antioxidant activity content [39]. Brewer spent grain solid state fermentation with *Rhizopus oligosporus* increased citric acid amount, amino acids level, antioxidant activity, and vitamin content. Compounds such as 3-Hydroxyanthranilic acid and myo-inositol were two important compounds identified during fermentation that already claimed to have anti-inflammatory and neuro-protective effects [40].

Two mesophilic solid-state fermentations of brewer spent grain were successfully used to enrich the nitrogen content and suggested there is a strong connection between the native microbiota and the inoculated one. A combination of yeasts (*Pichia kurdiavzevii*, *Saccharomyces cerevisiae*, *Kluyveromyces marxianus*), lactic acid bacteria (*Enterococcus faecium*, *Latilactobacillus graminis*, *Pediococcus acidilactici*), filamentous fungus (*Beauveria bassiana*), and *Gluconobacter diazotrophicus* revealed that the inoculated microbiota was unable to persist after 7–14 days during solid-state fermentation but was able to generate positive impact on the brewer spent grain nitrogen content [41].

In the fermentation process of rice bran with *R. oryzae* free phenolic content increased from 2.4 to 5.1 mg/g, while mixed fermentation of rice bran with *Monascus purpureus* and *Rhizopus oligosporus* strains increased with twofold higher the amount of phenolic compounds [42]. The rice bran solid-state fermentation with *Rhizopus oryzae* CCT 7560 increased after 48 hours of fermentation, the ash content from 1.1 to 56%, fiber content from 11 to 57%, and protein content from 6.1 to 49% [43]. The bran wheat fermentation with *Sacharomyces cerevisiae* increased the solubility of dietary fiber, protein, and ash and decreased the phytic acid content. It is important to mention that a key role in the wheat bran fermentation process was the size particle of the bran [32].

A big disadvantage of the fermentation process claimed by the literature is the long time process to achieve the desired bioactive compounds concentration together with the use of batch reactors in which the undesirable microorganism could develop [44]. Moreover, the cost needed for stirring or aeration during the conventional fermentation process is relatively high [45]. On the other side, the same issues were identified in solid-state fermentation, mentioning that there is less energy input due to the lower liquid phase and less operational costs such as aeration or stirring but there are some difficulties in monitoring temperature, humidity, and pH values [45].

To overcome this challenge, the extrusion process is claimed by a large body of literature as being a technological solution that combines short time, high temperature with pressure under a shear force, triggering molecular disruption, and leading to textural and structural changes in the raw materials [44, 46].

A hypothesis was formulated suggesting that the solubility of dietary fiber might be enhanced through extrusion cooking, likely attributed to the decrease in particle sizes of extruded samples and/or the disruption of covalent and non-covalent bonds in larger molecules resulting from the application of high temperatures [11]. This is in line with the finding of Grasso [47] who mentioned that the extrusion process could alter the molecular fiber structure and increase the soluble fiber amount and with Aktas-Akyildiz et al., [48] who showed that extrusion parameters are highly important in the increment of wheat bran fiber solubility together with enzymatic hydrolysis that could successfully further extend the dietary fiber solubility.

Accordingly, Haghighi-Manesh et al., [49] showed that extrusion and enzymatic treatments have a synergetic effect leading to a higher corn bran soluble dietary fiber. In line with this, extrusion process parameters (temperature 130°C, feed moisture 24%, and screw speed 400 rpm) of wheat bran increased the extractability of total dietary fiber with 40%, whilst arabinoxylan extractability increased from 5.8 to 9% [44]. Rice bran extrusion (temperature 100°C, feed moisture 60%, and 100 rpm) led to a total increase content of soluble fibers with 30% [44].

On the other side, extrusion cooking process enhanced the release of phenolic compounds and thermal processing of maize bran (160°C for 1 hour) allowed the solubilization of ferulic acid in a percentage of 80%, in the form of feruloylated oligosaccharides, that has higher antioxidant activity than free ferulic acid [50]. During extrusion of wheat bran feed moisture of 30°C followed by temperatures of 140 and 180°C highly improved the extraction yield of bound phenolic content [44]. The extrusion process of rice bran increased the phenolic bioaccessibility from 395.11 to 555.27 mg GAE/100 g, respectively [42].

The rice bran extrusion process improved its color, decreased the amount of phytic acid, and did not change the content of vitamins such as folic acid, niacin, pantothenic, or riboflavin amount. Likewise, a positive effect was observed in the amount of dietary fiber, which is involved in human constipation improvement [51]. Moreover, the use of rice bran in the starch-based extruded snacks is a tool for the improvement of the porous structure, increases the water holding capacity, retarded the starch retrogradation, and enhances the syneresis during storage of rice bran extruded snacks [52].

Another key role of the extrusion process is the capacity to inactivate enzymes such as lipases and lipoxygenases and therefore inhibit the cereal by-products post-milling rancidification [44].

4. Innovative applications in bakery, pastry, and gastronomy

Aspects of concern which have been investigated in various studies include the physical characteristics of by-products and the methods employed for their integration into new products. This involves preserving product stability, textural characteristics, and consumer acceptance while minimizing any adverse impact on the properties of the targeted compounds found in the cereal by-products. Therefore, the valorization of cereal by-products impact on the food chain with a focus on bakery, pastry, and gastronomy products will be further discussed in the present chapter.

The utilization of corn bran in the snack bar manufacturing was mentioned by Sousa et al., [53], who produced snack bars with a high percentage addition of corn bran (55%), resulting in a product with increased nutritional content in total dietary fiber (11.30 g/100 g), lower in carbohydrates (70.27 g/100 g) compared with the control (79.43 g/100 g) and higher in lipids (2.64 g/100 g). From the physical parameters point of view, the use of corn grain in a percentage of 55% influenced the increase of hardness value, from a value of 85.74 N (control sample) to a value of 126.26 N, respectively [53]. In this line, Grasso [47] showed that corn bran could be successfully used in extruded snacks manufacturing, with an inclusion range between 10 and 80%, resulting in enriched fiber corn extrudates.

Another study highlighted that the use of corn bran in cake batter manufacturing could be successful to a percentage addition of 20%, without negative effect on hardness, springiness, and accepted by consumers form the texture, taste, and overall acceptability attributes [13]. Sharma et al., [54] mentioned that blending of corn bran, defatted corn germ, and corn gluten with wheat flour increases the protein, minerals (Ca, P, and Fe), and crude fiber content and the effect of blending in bread, muffins, cookies, noodles, and extruded snacks is presented in **Table 1**.

Milled corn bran was previously submitted to an integrated extrusion-enzymatic treatment and afterward, included in a functional cake manufacturing process. An addition of 30% pretreated corn bran was considered too much for the panelist's textural and taste points of view, meantime, a 10% addition was the optimum percentage [49] that accomplished the general panelist expectations.

The use of wheat bran fermented with yeast and its 20% addition in bread manufacturing process positively influenced the protein (21.8%), soluble arabinoxylans (1.6%), and ash (6.6%) contents, while the enzyme-assisted yeast fermentation of bread wheat mainly improved soluble arabinoxylans content (3.8%) [33]. Tafton bread is a consumed product in Iran [32] and wheat bran fermented by *Sacharomyces cerevisiae* could be used at a replacement level of 15%, leading to an increase in bread shelf-life and freshness preservation [32].

The addition of 7.08% rice bran fermented with *Sacharomyces cerevisiae* strain in the gluten-free muffins preparation enhanced the increased of protein content from 10.5 to 12.7%, a decrease of the lipid from 33.5 to 26.7%, whilst, vanillin increased its value from 3.12 to 5.49 mg/g [57]. The previous fermentation of wheat bran with *Basidiomycete* BS-01 strain was able to diminish the negative effect of bran addition on the steamed bread quality improving the dough viscoelasticity and bread-specific volume [56].

Spent malt rootlets fermentation with different lactic acid bacteria such as *L. citreum* TR116, *W. cibaria* MG1, *L. plantarum* FST1.7, *L. reuteri* R29, and *L. amylovorus* FST2.11, improved the breadcrumb texture, hardness and resilience of the final baked steamed bread and some antifungal natural compounds such as hydroxyphenyllactic acid and 4-hydroxybenzoic acid were formed, increasing the shelf-life of the bread. Also, the sensory characteristics of the bread were enhanced, resulting in a flavored bread, which can be compared to that of a sourdough conventional bread [21].

Lactobacillus plantarum f10 and *Lactobacillus rhamnosus* GG were successfully used for the milled and spray-dried brewer spent grain fermentation leading to an improvement of the dough characteristics such as gluten development, dough resistance/stiffness, and enhanced properties of the final bread [59], as presented in **Table 1**.

Biscuits are considered popular products which nowadays are nutritionally deficient in some nutrients such as minerals, protein, and even dietary fiber [60]. The utilization of a 15% addition of spent malt rootlets had a positive influence on the sensorial and nutritional composition of biscuits such as minerals, fatty acid content, and highly improved the aroma volatile compounds of the baked goods [60].

It is worthy of note that a combination of fermentation treatment together with an enzymatic one could be the best solution for improving textural characteristics, loaf volume and enhancing the shelf life of the bread enriched with wheat bran [75].

The utilization of brewer spent grain in breadsticks manufacturing was mentioned by Ktenioudaki et al., [63] as displayed in **Table 1**, and its utilization in bread

By-products	New product	Maximum addition	Improvement on the final product
		Bakery and pastry products	
Corn bran	Snack bars	55%	↑ of dietary fiber, ↑ lipid content, and ↑ mineral content, [53]
Corn bar	Extruded snacks	10–80%	↑ total dietary fiber content [47]
Corn bran	Cake batters	20%	↑ dietary fiber, hardness and springiness not affected, and accepted by consumers [13]
Corn bran, defatted germ and gluten	Noodles and extruded snacks	10%	↑ cooking time; ↑ water absorption ratio; and gluten blended snacks ↑ overall acceptability
Corn bran, defatted germ and gluten	Bread, cookies, and muffins	5–10%	↑ volume bread made with corn gluten; corn germ muffins ↑ acceptability scores;
Fermented wheat bran	Wheat bran	20%	↑ fiber (arabinoxylans) solubilization [33]
Fermented wheat bran and enzyme treatment	Wheat bran	20%	Hardness and volume improved [33]
Wheat bran	Extrudates	10–20%	↑ soluble dietary fiber; ↓ insoluble fiber content [47]
Corn bran	Functional cake	5–10%	↑ soluble dietary fiber; ↑ desirable sensorial properties; and ↑ textural properties (↓cohesiveness and springiness, ↑ gumminess), [49]
Wheat bran	Tafton bread	15%	↑ soluble dietary fiber; ↑ freshness and shelf-life [32]
Wheat bran pre-fermented and enzyme treatment	Bread		↑ arabinoxylan solubility in the dough; ↑dough rheological characteristics; and ↑quality of bread [55]
Wheat bran fermented	Steamed bread	3–9%	↑ bread resilience; ↑bread specific volume; [56]
Rice bran fermented	Gluten-free cookies	7.08%	↑ protein level; ↑phenolic compounds; ↑ antioxidant activity; and ↑shelf-life [57].
Defatted rice bran	Bread	5–10%	↑ dietary fiber content; ↑ antioxidant activity and shelf-life [58]
Spent malt rootlets fermented	Steamed bread	5%	↑ specific volume; ↑crumb softness; and ↑ bread flavors [21].
Spent malt rootlets fermented	Bread	15%	↑ specific volume; ↓ crumb hardness; limited microbial growth rate; and ↓ the release of sugar during digestion [59]
Spent malt rootles	Multigrain biscuits	15%	↑ fatty acid content; ↑ volatile compound; and ↑ sensorial characteristics [60]
Brewer spent grain	Muffins	6%	↑ antioxidant activity; ↓ peroxide and thiobarbituric acid [61]
Brewer spent grain	Bread	10%	↑ protein, mineral, and dietary fiber [62]
Brewer spent grain	Breadsticks	15–35%	↑ dietary fiber content and protein; ↓ crispiness; and ↓ lower baking volume [63]

By-products	New product	Maximum addition	Improvement on the final product
		Bakery and pastry products	
Brewer spent grain	Bread	20–25%	↑ phenolic content; ↑ soluble and insoluble fiber; ↓ bread texture; and ↑ hardness [64]
Brewer spent grain	Bread	5–20%	↑ mineral, protein, fiber, and fat [65].
Brewer spent grain	Dry pasta	10 g/100 g	↑ protein, dietary fiber, antioxidant activity, and β-Glucan [66]
		Gastronomy products	
Wheat bran	Chicken meat biscuits	5%	↑ oxidative stability; ↑ crude fiber and ash contents [67]
Wheat bran extract	Beef meat hamburgers	0.8%	↓ the formation of oxidation products; ↑antioxidant potential [68]
Wheat bran	Meatballs	5–20%	↓ total fat content; ↓ total *trans* fatty acids; ↑ ash, protein; and ↓ moisture [69]
Corn bran	Turkish meatballs	10–15%	↓ weight losses; ↑ protein content; and ↑ fiber amount [70]
Brewer spent grain	Bratwurst sausage	3%	↑ protein content; ↑ color, chewiness, appearance, and pastiness [71]
Brewer spent grain	Chicken patties	3%	↑ dietary fiber; improve the cooking loss [72]
Brewer spent grain	Smoked sausages	3%	↑protein, ash; ↑ freshness [73]
Brewer spent grain	Beef frankfurters	1–5%	↑ total dietary fiber [74]
Spent malt rootles	Sausage	10%	↓ cooking losses; ↑ fiber content [22]

↑ increase; ↓ decrease.

Table 1.
Effects of by-products cereal addition in different products.

manufacturing was mentioned by Fărcaș et al., [65] and Baiano et al., [64]. The use of brewer spent grain previously fermented with *Lactobacillus plantarum* FST 1.7. strain leads to an increase in the nutritional parameters but also in softer breads with improved springiness [62]. It is worthy to mention the use of brewer spent grain in the pasta manufacturing, by replacing semolina flour with 10 g of brewer spent grain [66]. The utilization of 6% addition of brewer spent grain hydrolysate in muffins manufacturing increased radical scavenging activity six times higher compared to the control sample without any hydrolysate addition and increased α- glucosidase and α-amylase inhibition with values of 40 and 88%, respectively [61]. With respect to the sensory attributes, an addition of 2% brewer spent grain hydrolysate in muffins manufacturing obtained a similar score for overall acceptance, color, texture, flavor, and general aspects as the control sample [61].

On the other side, the utilization of the selected by-products in gastronomy products is displayed in **Table 1**. The use of wheat bran extract in a percentage of 0.8% in beef meat hamburgers significantly increases the global antioxidant response and the product oxidation stability [68] with possible benefits on human health. It is worth to mention also that wheat bran was used in percentages of 5, 10, 15, and 20% in low-fat meatballs and highly improved the weight loss, the nutritional content of the meatballs (protein, ash), color attributes, decreased the salt content, and registered the lowest moisture. From the sensorial characteristics, the samples with 20% wheat bran addition encoded the biggest firmness, probably because the addition of wheat bran decreased total fat content [69].

It is worth mentioning that these by-products could be successfully used in gastronomy, specifically in the meat manufacturing products. For instance, the utilization of brewer spent grain in the meat products was mentioned by Talens et al., [71] who demonstrated that brewer spent grain could be used as a source of protein in the hybrid sausages, whilst, Kim et al., [72] used brewer spent grain in the chicken patties in a percentage addition of 3%. Corn bran improves the fiber of cooked Turkish meatballs and decreases their fat content [70], meantime, a 3% addition of brewer spent grain does not affect the textural and sensorial characteristics of the smoked sausages [73]. In the beef frankfurters manufacturing, brewer spent grain could be used as a fat substitute for produce products with high dietary fiber and low-fat content [74].

5. Health impact of cereal by-products consumption

The main positive effects of cereal by-products consumption on human health are presented in **Figure 3**. The major antioxidants from rice bran are γ-oryzanol, phytosterol from which campesterol, sitosterol, and stigmasterol are the major ones, and tocols (tocotrienol and tocopherol). Tocols, also entitled vitamin E are known to be natural antioxidants with positive effects on certain cancer cells and to reduce the cholesterol concentration, [76] as shown in **Table 2**. Rice bran is also a rich source of B-complex vitamins, carotenoids, phenolic compounds, octacosanol, and also squalene [82]. It is worthy of note that in rice bran and its products, more than 100 antioxidant compounds were identified from which 12 phenolic acid and several flavonoids such as tricin, quercetin, luteolin, kaempferol, and apigenin [82].

Cyanidin-3-O-glucoside and peonidin-3-O-glucoside were the main compounds identified in glutinous black rice varieties with higher anti-inflammatory and

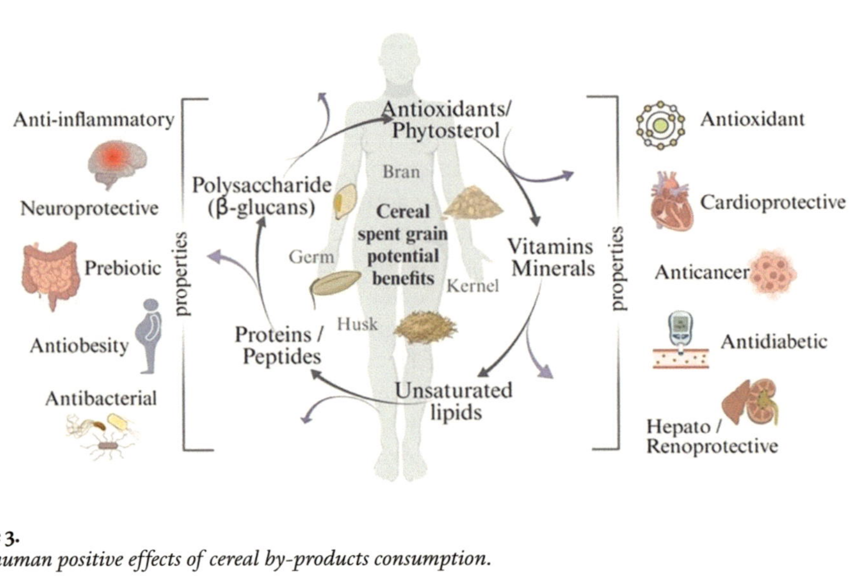

Figure 3.
Main human positive effects of cereal by-products consumption.

Cereal by-products	Bioactive compound	Beneficial properties
Rice bran	Bioactive peptides	control of hypertension, oxidative stress, type 2 diabetes mellitus, and various abnormal cellular processes [77].
Wheat germ	Wheat germ extract	fermented wheat germ extracts beneficial effect against ovarian cancer with anti-inflammatory and immunomodulatory effects [78].
Wheat bran	Arabinoxylan oligosaccharides	Improve glucose tolerance, insulin sensitivity mechanism, and gut fermentation improved [79].
Wheat bran	Arabinoxylans and β-glucans	Play a key role in reducing the susceptibility to type II diabetes, colorectal cancer, as well as cardiovascular and diverticular diseases, [29]
Wheat bran	Phenolic acid Ferulic acid	↑ Antioxidant properties, protection against heart diseases and cancer [80]. Cardiovascular diseases, chronic diseases [31].
Corn bran	Arabinoxylans	Prebiotic effect, improve bowel movement, improve the development of colonic microbiota, and the production of exopolysaccharides [5]. Improve gut microbiota modulation [81]
Corn bran	Dietary fiber	52 g/day reduce triglyceride content, glycosylated hemoglobin, and very low-density lipoprotein cholesterol [12]; ↑ fecal bulk, ↓ gastrointestinal transit time, and ↑ colon health [12]
Corn bran	Hemicellulose	10 g/day for 6 months improve obese individual's glycemic and insulinemic responses [12]
Rice bran	γ-oryzanol	Antioxidant activity, lipid-lowering effect, antidiabetic effect, and anti-cancer properties; improve kidney and liver functions in rats [15]
Rice bran	γ-oryzanol, tocols, phytosterols	Antioxidant, hypocholesterolemia effects; ↓ serum cholesterol concentration [76]
Rice bran	Polysaccharides	Weight reduction, antihypertensive, and hypolipidemic activity [15]
Rice bran	Dietary fiber	Prebiotic characteristics [15]
Rice bran	Phenolic acid and flavonoids	Antioxidant and antimicrobial effect [82]
Rice bran	Arabinoxylan and β-glucan	Functional polysaccharides with immunomodulatory effects on mammary tumors, gastrointestinal cancers, and Lewis lung carcinomas [82]
Rice bran	Oryzatensin	Bioactive peptide with immunomodulatory effects [82]
Black bran extracts	Cyanidin-3-O-glucoside Peonidin-3-O-glucoside	↑ antioxidant and anti-inflammatory properties [14]
Brewer spent grain	Phenolic compounds	↑ Antioxidant activity, immunomodulatory effect; antimutagenic effects and DNA-protective [17]
Brewer spent grain	β-Glucan	↑ colon functionality [82] ↓ plasma cholesterol [64]
Brewer spent grain	Peptides	Thrombin inhibitory activity [83]

Cereal by-products	Bioactive compound	Beneficial properties
Brewer spent grain	Arabinoxylans	↓ serum cholesterol; ↑ calcium and magnesium adsorption; antioxidant properties [64]
Brewer spent grain	Ferulic acid	Antioxidant, anti-inflammatory, and antimicrobial [36]
Brewer spent grain	Protein hydrolysate	Antioxidant and antidiabetic effects [61]
Spent malt rootlets	Protein hydrolysates	↑ antihypertensive activity [84]

↑ increase; ↓ decrease.

Table 2.
Human benefits of cereal by-products bioactive compounds.

antioxidant effects than non-glutinous rice varieties [14]. The oral administration of a formulation comprising powdered black rice and bran among a group of healthy individuals aged between 65 and 75 years resulted in a reduction in the levels of the inflammatory marker IL-6 and C-reactive protein [14].

In a study where hemicellulose was extracted from corn bran, rice bran, and wheat bran, using the same extraction method, and submitted to human fecal microflora for the *in vitro* fermentation, revealed that corn bran hemicellulose exhibited higher fermentability, yielding significantly greater quantities of acetate, and propionate compared to hemicellulose from rice and wheat bran. These variations can be attributed to differences in their monosaccharide compositions, derivative-unit distributions, and molecular weights.

The effective fermentation of corn bran hemicellulose, leading to the production of short-chain fatty acids (SCFAs), suggests its potential to contribute to colon health benefits upon consumption [12].

Brewer spent grain β-glucan molar mass is highly influenced by the proteolytic enzymes of the gastrointestinal tract, enhancing its functional role in the colon [83]. Briefly, it seems that proteins could be linked to β-glucan and therefore could affect the molar mass where a key role in the β-glucan hydrodynamic size is played by the enzymatic treatment [85]. Peptides hydrolysates of brewer spent grain are able to inhibit coagulation factors in the blood coagulation pathway and inhibit the thrombin action, thereby preventing the fibrin conversion to fibrinogen [83].

Protein hydrolysates extracts from spent malt rootlets grain treated with different enzymes showed a high angiotensin-converting enzyme (ACE) inhibitory activity and could also have an antihypertensive effect [84].

6. Conclusion

The ever-increasing demand for healthy food correlated with the need of a sustainable future economy, gained researchers' attention, and increased the utilization of by-products in food manufacturing process. Cereal by-products are rich sources of several bioactive compounds such as dietary fiber, protein, amino acids, minerals, phenolic acids, flavonoids, and vitamins. Unfortunately, there are several antinutrients factors and macromolecular complexes (such as that of protein and fiber) that influence in a negative way the bioaccesibility and biodisponibility of the aforementioned bioactive compounds. Moreover, from the technological point of view, some cereal by-products lead to undesirable texture profiles of the final products such as increased hardness and cohesiveness values and a decrease in gumminess and chewiness attributes.

A way to tackle this drawback is the use of pretreatments such as submerged fermentation or solid-state fermentation of cereal by-products with different lactic acid bacteria strains or even enzymatic treatments. Moreover, extrusion is a new tool that could improve the fiber solubility with no negative effects on other bioactive compounds. The further valorization of these by-products in bakery, pastry, and gastronomy products represents a shift toward a circular food economy system and improves the nutritional value of the final products. This chapter is an important step in the journey toward a new possibility to valorize cereal by-product.

Acknowledgements

This research was supported by a grant of the Romanian Ministry of Education and Research, CCCDI-UEFISCDI, project number PN-III-P2-2.1-PED-2019-3622, within PNCDI III.

Author details

Maria Simona Chiș and Anca Corina Fărcaș*
Faculty of Food Science and Technology, University of Agricultural Sciences and Veterinary Medicine Cluj-Napoca, Cluj-Napoca, Romania

*Address all correspondence to: anca.farcas@usamvcluj.ro

References

[1] Guo H, Wu H, Sajid A, Li Z. Whole grain cereals: The potential roles of functional components in human health. Critical Reviews in Food Science and Nutrition. 2022;**62**(30):8388-8402. DOI: 10.1080/10408398.2021.1928596

[2] Papageorgiou M, Skendi A. In: Galanakis CM, editor. Sustainable Recovery and Reutilization of Cereal Processing by-Products Introduction to Cereal Processing and by-Products. Sawston, United Kingdom: Woodhead Publishing Series in Food Science, Technology and Nutrition, Elsevier; 2018. pp. 1-25. DOI: 10.1016/B978-0-08-102162-0.00001-0

[3] Luithui Y, Baghya Nisha R, Meera MS. Cereal by-products as an important functional ingredient: Effect of processing. Journal of Food Science and Technology. 2019;**56**(1):1-11. DOI: 10.1007/s13197-018-3461-y

[4] Arzami AN, Ho TM, Mikkonen KS. Valorization of cereal by-product hemicelluloses: Fractionation and purity considerations. Food Research International. 2022;**151**:1-21. DOI: 10.1016/j.foodres.2021.110818

[5] Fărcaș AC, Socaci SA, Nemeș SA, Pop OL, Coldea TE, Fogarasi M, et al. An update regarding the bioactive compound of cereal by-products: Health benefits and potential applications. Nutrients. 2022;**14**(17):1-19. DOI: 10.3390/nu14173470

[6] Salazar-lópez NJ, Ovando-martínez M, Domínguez-avila JA. Cereal/grain by-products. In: Campos-Vega R, Dave Oomah B, Vergara-Castañeda HA, editors. Food Wastes and by-Products: Nutraceutical and Health Potential. John Wiley & Sons Ltd; 2020. pp. 1-34. DOI: 10.1002/9781119534167

[7] Fărcaș A, Drețcanu G, Pop TD, Enaru B, Socaci S, Diaconeasa Z. Cereal processing by-products as rich sources of phenolic compounds and their potential bioactivities. Nutrients. 2021;**13**(11):1-17. DOI: 10.3390/nu13113934

[8] Alabiden Tlais AZ, Fiorino GM, Polo A, Filannino P, Di Cagno R. High-value compounds in fruit, vegetable and cereal byproducts: An overview of potential sustainable reuse and exploitation. Molecules. 2020;**25**:1-27. DOI: 10.3390/molecules25132987

[9] Koistinen VM, Tuomainen M, Lehtinen P, Peltola P, Auriola S, Jonsson K, et al. Side-stream products of malting: A neglected source of phytochemicals. npj Science of Food. 2020;**4**(1):1-9. DOI: 10.1038/s41538-020-00081-0

[10] Roth M, Jekle M, Becker T. Opportunities for upcycling cereal byproducts with special focus on Distiller's grains. Trends in Food Science and Technology. 2019;**91**:282-293. DOI: 10.1016/j.tifs.2019.07.041

[11] Skendi A, Zinoviadou KG, Papageorgiou M, Rocha JM. Advances on the valorisation and functionalization of by-products and wastes from cereal-based processing industry. Food. 2020;**9**:1-28. DOI: 10.3390/foods9091243

[12] Ai Y, Jane JL. Macronutrients in corn and human nutrition. Comprehensive Reviews in Food Science and Food Safety. 2016;**15**(3):581-598. DOI: 10.1111/1541-4337.12192

[13] Singh M, Liu SX, Vaughn SF. Effect of corn bran as dietary fiber addition on baking and sensory quality. Biocatalysis and Agricultural Biotechnology.

2012;**1**(4):348-352. DOI: 10.1016/j.bcab.2012.02.005

[14] Mapoung S, Semmarath W, Arjsri P, Thippraphan P, Srisawad K, Umsumarng S, et al. Comparative analysis of bioactive-phytochemical characteristics, antioxidants activities, and anti-inflammatory properties of selected black rice germ and bran (*Oryza sativa* L.) varieties. European Food Research and Technology. 2023;**249**(2):451-464. DOI: 10.1007/s00217-022-04129-1

[15] Sapwarobol S, Saphyakhajorn W, Astina J. Biological functions and activities of Rice bran as a functional ingredient: A review. Nutrition and Metabolic Insights. 2021;**14**:1-11. DOI: 10.1177/11786388211058559

[16] Aggelopoulos T, Bekatorou A, Pandey A, Kanellaki M, Koutinas AA. Discarded oranges and brewer's spent grains as promoting ingredients for microbial growth by submerged and solid state fermentation of agro-industrial waste mixtures. Applied Biochemistry and Biotechnology. 2013;**170**(8):1885-1895. DOI: 10.1007/s12010-013-0313-0

[17] Naibaho J, Korzeniowska M. Brewers' spent grain in food systems: Processing and final products quality as a function of fiber modification treatment. Journal of Food Science. 2021;**86**(5):1532-1551. DOI: 10.1111/1750-3841.15714

[18] Olivares-Galván S, Marina ML, García MC. Extraction of valuable compounds from brewing residues: Malt rootlets, spent hops, and spent yeast. Trends in Food Science and Technology. 2022;**127**:181-197. DOI: 10.1016/j.tifs.2022.06.002

[19] Fărcaș AC, Socaci SA, Chiș MS, Martí J. In vitro digestibility of minerals and B vitamins group from different brewers' spent grains. Nutrients. 2022;**27**(7):1-15. DOI: 10.3390/plants10112504

[20] Ludka FR, Klosowski AB, Camargo GA, Justo AS, Andrade EA, Beltrame FL, et al. Brewers' spent grain extract as antioxidants in starch-based active biopolymers. International Journal of Food Science and Technology. 2024;**59**(1):142-150. DOI: 10.1111/ijfs.16785

[21] Neylon E, Nyhan L, Zannini E, Sahin AW, Arendt EK. From waste to taste: Application of fermented spent rootlet ingredients in a bread system. Food. 2023;**12**(7):1-25. DOI: 10.3390/foods12071549

[22] Neylon E, Arendt EK, Lynch KM, Zannini E, Bazzoli P, Monin T, et al. Rootlets, a malting by-product with great potential. Fermentation. 2020;**6**(4):117. DOI: 10.3390/fermentation6040117

[23] Budaraju S, Mallikarjunan K, Annor G, Schoenfuss T, Raun R. Effect of pre-treatments on the antioxidant potential of phenolic extracts from barley malt rootlets. Food Chemistry. 2018;**266**:31-37. DOI: 10.1016/j.foodchem.2018.05.110

[24] Karlović A, Jurić A, Ćorić N, Habschied K, Krstanović V, Mastanjević K. By-products in the malting and brewing industries-re-usage possibilities. Fermentation. 2020;**6**(3):1-17. DOI: 10.3390/FERMENTATION6030082

[25] Pejin J, Radosavljević M, Pribić M, Kocić-Tanackov S, Mladenović D, Djukić-Vuković A, et al. Possibility of L-(+)-lactic acid fermentation using malting, brewing, and oil production by-products. Waste Management. 2018;**79**:153-163. DOI: 10.1016/j.wasman.2018.07.035

[26] Parenti O, Guerrini L, Zanoni B. Techniques and technologies for the breadmaking process with unrefined wheat flours. Trends in Food Science and Technology. 2020;**99**:152-166. DOI: 10.1016/j.tifs.2020.02.034

[27] Navrotskyi S, Guo G, Baenziger PS, Xu L, Rose DJ. Impact of wheat bran physical properties and chemical composition on whole grain flour mixing and baking properties. Journal of Cereal Science. 2019;**89**:102790. DOI: 10.1016/j.jcs.2019.102790

[28] Acin-Albiac M, Filannino P, Arora K, Da Ros A, Gobbetti M, Di Cagno R. Role of lactic acid bacteria phospho-β-glucosidases during the fermentation of cereal by-products. Food. 2021;**10**(1):1-14. DOI: 10.3390/foods10010097

[29] Verni M, Rizzello CG, Coda R. Fermentation biotechnology applied to cereal industry by-products: Nutritional and functional insights. Frontiers in Nutrition. 2019;**6**:1-13. DOI: 10.3389/fnut.2019.00042

[30] Spaggiari M, Ricci A, Calani L, Bresciani L, Neviani E, Dall'Asta C, et al. Solid state lactic acid fermentation: A strategy to improve wheat bran functionality. LWT. 2020;**118**:108668. DOI: 10.1016/j.lwt.2019.108668

[31] Deroover L, Tie Y, Verspreet J, Courtin CM, Verbeke K. Modifying wheat bran to improve its health benefits. Critical Reviews in Food Science and Nutrition. 2020;**60**(7):1104-1122. DOI: 10.1080/10408398.2018.1558394

[32] Rezaei S, Najafi MA, Haddadi T. Effect of fermentation process, wheat bran size and replacement level on some characteristics of wheat bran, dough, and high-fiber Tafton bread. Journal of Cereal Science. 2019;**85**:56-61. DOI: 10.1016/j.jcs.2018.11.019

[33] Hartikainen K, Poutanen K, Katina K. Influence of bioprocessed wheat bran on the physical and chemical properties of dough and on wheat bread texture. Cereal Chemistry. 2014;**91**(2):115-123. DOI: 10.1094/CCHEM-04-13-0074-R

[34] Bolarinwa IF, Hanis-Syazwani MG, Muhammad K. Optimisation of important processing conditions for rice bran sourdough fermentation using *Lactobacillus plantarum*. Foods and Raw Materials. 2019;**7**(1):131-142. DOI:10.21603/2308-4057-2019-1-131-142

[35] Neylon E, Nyhan L, Zannini E, Monin T, Münch S, Sahin AW, et al. Food ingredients for the future: In-depth analysis of the effects of lactic acid Bacteria fermentation on spent barley rootlets. Fermentation. 2023;**9**(1):1-21. DOI: 10.3390/fermentation9010078

[36] L-Shwafy KWA, Chadni M, Hariz Abg Zamari MH, Ioannou I. Enzymatic extraction of ferulic acid from brewer's spent grain: Effect of physical pretreatments and optimization using design of experiments. Biocatalysis and Agricultural Biotechnology. 2023;**51**:1-11. DOI: 10.1016/j.bcab.2023.102779

[37] Bazsefidpar N, Ahmadi Gavlighi H, Ghandehari Yazdi AP, Jafari SM. Optimization of protein extraction from brewer's spent grain and production of bioactive peptides. Biomass Convers Biorefinery. 2023:1-11. DOI: 10.1007/s13399-023-03932-4

[38] Roasa J, De Villa R, Mine Y, Tsao R. Phenolics of cereal, pulse and oilseed processing by-products and potential effects of solid-state fermentation on their bioaccessibility, bioavailability and health benefits: A review. Trends in Food Science and Technology. 2021;**116**:954-974. DOI: 10.1016/j.tifs.2021.08.027

[39] Cervantes-Ramirez JG, Vasquez-Lara F, Sanchez-Estrada A, Troncoso-Rojas R, Heredia-Olea E, Islas-Rubio AR. Arabinoxylans release from brewers' spent grain using extrusion and solid-state fermentation with *fusarium oxysporum* and the antioxidant capacity of the extracts. Food. 2022;**11**(10):1-15. DOI: 10.3390/foods11101415

[40] Cooray ST, Chen WN. Valorization of brewer's spent grain using fungi solid-state fermentation to enhance nutritional value. Journal of Functional Foods. 2018;**42**:85-94. DOI: 10.1016/j.jff.2017.12.027

[41] Bianco A, Zara G, Garau M, Castaldi P, Atzori AS, Deroma MA, et al. Microbial community assembly and chemical dynamics of raw brewers' spent grain during inoculated and spontaneous solid-state fermentation. Waste Management. 2024;**174**:518-527. DOI: 10.1016/j.wasman.2023.12.021

[42] Chen Y, Ma Y, Dong L, Jia X, Liu L, Huang F, et al. Extrusion and fungal fermentation change the profile and antioxidant activity of free and bound phenolics in rice bran together with the phenolic bioaccessibility. LWT. 2019;**115**:108461. DOI: 10.1016/j.lwt.2019.108461

[43] Adebo JA, Njobeh PB, Gbashi S, Oyedeji AB, Ogundele OM, Oyeyinka SA, et al. Fermentation of cereals and legumes: Impact on nutritional constituents and nutrient bioavailability. Fermentation. 2022;**8**(2):1-57. DOI: 10.3390/fermentation8020063

[44] Orozco-Angelino X, Espinosa-Ramírez J, Serna-Saldívar SO. Extrusion as a tool to enhance the nutritional and bioactive potential of cereal and legume by-products. Food Research International. 2023;**169**:1-13. DOI: 10.1016/j.foodres.2023.112889

[45] Sun X, Dou Z, Shurson GC, Hu B. Bioprocessing to upcycle agro-industrial and food wastes into high-nutritional value animal feed for sustainable food and agriculture systems. Resources, Conservation and Recycling. 2024;**201**:1-13. DOI: 10.1016/j.resconrec.2023.107325

[46] Igual M, García-Segovia P, Martínez-Monzó J. Effect of *Acheta domesticus* (house cricket) addition on protein content, colour, texture, and extrusion parameters of extruded products. Journal of Food Engineering. 2020;**282**:1-7. DOI: 10.1016/j.jfoodeng.2020.110032

[47] Grasso S. Extruded snacks from industrial by-products: A review. Trends in Food Science and Technology. 2020;**99**:284-294. DOI: 10.1016/j.tifs.2020.03.012

[48] Aktas-Akyildiz E, Masatcioglu MT, Köksel H. Effect of extrusion treatment on enzymatic hydrolysis of wheat bran. Journal of Cereal Science. 2020;**93**(8):1-7. DOI: 10.1016/j.jcs.2020.102941

[49] Haghighi-Manesh S, Azizi MH. Integrated extrusion-enzymatic treatment of corn bran for production of functional cake. Food Science & Nutrition. 2018;**6**(7):1870-1878. DOI: 10.1002/fsn3.738

[50] Santos D, Pintado M, Lopes da Silva JA. Potential nutritional and functional improvement of extruded breakfast cereals based on incorporation of fruit and vegetable by-products – A review. Trends in Food Science and Technology. 2022;**125**:136-153. DOI: 10.1016/j.tifs.2022.05.010

[51] Rafe A, Sadeghian A, Hoseini-Yazdi SZ. Physicochemical, functional, and nutritional characteristics of stabilized rice bran form tarom cultivar. Food Science

& Nutrition. 2017;**5**(3):407-414. DOI: 10.1002/fsn3.407

[52] Wang P, Fu Y, Wang L, Saleh ASM, Cao H, Xiao Z. Effect of enrichment with stabilized rice bran and extrusion process on gelatinization and retrogradation properties of rice starch. Starch/Staerke. 2017;**69**(7-8). DOI: 10.1002/star.201600201

[53] Sousa MF d, Guimarães RM, Araújo M d O, Barcelos KR, Carneiro NS, Lima DS, et al. Characterization of corn (*Zea mays* L.) bran as a new food ingredient for snack bars. LWT. 2019;**101**:812-888. DOI: 10.1016/j.lwt.2018.11.088

[54] Sharma S, Gupta JP, Nagi HPS, Kumar R. Effect of incorporation of corn byproducts on quality of baked and extruded products from wheat flour and semolina. Journal of Food Science and Technology. 2012;**49**(5):580-600. DOI: 10.1007/s13197-011-0304-5

[55] Messia MC, Reale A, Maiuro L, Candigliota T, Sorrentino E, Marconi E. Effects of pre-fermented wheat bran on dough and bread characteristics. Journal of Cereal Science. 2016;**69**:138-144. DOI: 10.1016/j.jcs.2016.03.004

[56] Li Z, Wang Z, Li LM, Zheng XL, Ma S. Effects of fermented wheat bran on flour, dough, and steamed bread characteristics. Journal of Chemistry. 2018;**51**:1-7. DOI: 10.1155/2018/1597308

[57] Christ-Ribeiro A, Chiattoni LM, Mafaldo CRF, Badiale-Furlong E, Souza-Soares LA d. Fermented rice-bran by *Saccharomyces cerevisiae*: Nutritious ingredient in the formulation of gluten-free cookies. Food Bioscience. 2021;**40**:100859. DOI: 10.1016/j.fbio.2020.100859

[58] Sairam S, Gopala Krishna AG, Urooj A. Physico-chemical characteristics of defatted rice bran and its utilization in a bakery product. Journal of Food Science and Technology. 2011;**48**(4):478-483. DOI: 10.1007/s13197-011-0262-y

[59] Neylon E, Arendt EK, Zannini E, Sahin AW. Fermentation as a tool to revitalise brewers' spent grain and elevate techno-functional properties and nutritional value in high fibre bread. Food. 2021;**10**(7):1-20. DOI: 10.3390/foods10071639

[60] Chiş MS, Pop A, Păucean A, Socaci SA, Alexa E, Man SM, et al. Fatty acids, volatile and sensory profile of multigrain biscuits enriched with spent malt rootles. Molecules. 2020;**25**(3):1-17. DOI: 10.3390/molecules25030442

[61] Bazsefidpar N, Ghandehari Yazdi AP, Karimi A, Yahyavi M, Amini M, Ahmadi Gavlighi H, et al. Brewers spent grain protein hydrolysate as a functional ingredient for muffins: Antioxidant, antidiabetic, and sensory evaluation. Food Chemistry. 2024;**435**:1-6. DOI: 10.1016/j.foodchem.2023.137565

[62] Waters DM, Jacob F, Titze J, Arendt EK, Zannini E. Fibre, protein and mineral fortification of wheat bread through milled and fermented brewer's spent grain enrichment. European Food Research and Technology. 2012;**235**(5):767-778. DOI: 10.1007/s00217-012-1805-9

[63] Ktenioudaki A, Chaurin V, Reis SF, Gallagher E. Brewer's spent grain as a functional ingredient for breadsticks. International Journal of Food Science and Technology. 2012;**47**(8):1765-1771. DOI: 10.1111/j.1365-2621.2012.03032.x

[64] Baiano A, la Gatta B, Rutigliano M, Fiore A. Functional bread produced in a circular economy perspective: The use of brewers' spent grain. Food. 2023;**12**(4):1-27. DOI: 10.3390/foods12040834

[65] Fărcaş AC, Socaci SA, Tofană M, Mureşan C, Mudura E, Salanţă L, et al. Nutritional properties and volatile profile of brewer's spent grain supplemented bread. Romanian Biotechnology Letters. 2014;**19**(5):9705-9714

[66] Nocente F, Taddei F, Galassi E, Gazza L. Upcycling of brewers' spent grain by production of dry pasta with higher nutritional potential. LWT. 2019;**114**:1-6. DOI: 10.1016/j.lwt.2019.108421

[67] Kumar P, Chatli MK, Mehta N, Malav OP, Verma AK, Kumar D. Quality attributes and storage stability of chicken meat biscuits incorporated with wheat and oat bran. Journal of Food Quality. 2016;**39**(6):649-657. DOI: 10.1111/jfq.12232

[68] Šulniute V, Jaime I, Rovira J, Venskutonis PR. Rye and wheat bran extracts isolated with pressurized solvents increase oxidative stability and antioxidant potential of beef meat hamburgers. Journal of Food Science. 2016;**81**(2):H519-H527. DOI: 10.1111/1750-3841.13197

[69] Yilmaz I. Physicochemical and sensory characteristics of low fat meatballs with added wheat bran. Journal of Food Engineering. 2005;**69**(3):369-373. DOI: 10.1016/j.jfoodeng.2004.08.028

[70] Yasarlar EE, Daglioglu O, Yilmaz I. Effects of cereal bran addition on chemical composition, cooking characteristics and sensory properties of Turkish meatballs. Asian Journal of Chemistry. 2007;**19**(3):2353-2361

[71] Talens C, Llorente R, Simó-Boyle L, Odriozola-Serrano I, Tueros I, Ibargüen M. Hybrid sausages: Modelling the effect of partial meat replacement with broccoli, upcycled Brewer's spent grain and insect flours. Food. 2022;**11**(21):1-18. DOI: 10.3390/foods11213396

[72] Kim HW, Hwang KE, Song DH, Lee SY, Choi MS, Bin LY, et al. Effects of dietary fiber extracts from brewer's spent grain on quality characteristics of chicken patties cooked in convective oven. Korean Journal for Food Science of Animal Resources. 2013;**33**(1):45-52. DOI: 10.5851/kosfa.2013.33.1.45

[73] Nagy M, Semeniuc CA, Socaci SA, Pop CR, Rotar AM, Sălăgean CD, et al. Utilization of brewer's spent grain and mushrooms in fortification of smoked sausages. Food Science and Technology. 2017;**37**(2):315-320. DOI: 10.1590/1678-457X.23816

[74] Özvural EB, Vural H, Gökbulut I, Özboy-Özbaş Ö. Utilization of brewer's spent grain in the production of frankfurters. International Journal of Food Science and Technology. 2009;**44**(6):1093-1099. DOI: 10.1111/j.1365-2621.2009.01921.x

[75] Hemdane S, Jacobs PJ, Dornez E, Verspreet J, Delcour JA, Courtin CM. Wheat (*Triticum aestivum* L.) bran in bread making: A critical review. Comprehensive Reviews in Food Science and Food Safety. 2016;**15**(1):28-42. DOI: 10.1111/1541-4337.12176

[76] Raghav PK, Agarwal N, Sharma A. Emerging health benefits of rice bran – A review. IJMRME. 2016;**2**(1):367-382

[77] Udenigwe CC. Towards rice bran protein utilization: In silico insight on the role of oryzacystatins in biologically-active peptide production. Food Chemistry. 2016;**191**:135-138. DOI: 10.1016/j.foodchem.2015.01.043

[78] Koh EM, Lee EK, Song J, Kim SJ, Song CH, Seo Y, et al. Anticancer activity and mechanism of action of fermented

wheat germ extract against ovarian cancer. Journal of Food Biochemistry. 2018;**42**(6):1-11. DOI: 10.1111/jfbc.12688

[79] Boll EVJ, Ekström LMNK, Courtin CM, Delcour JA, Nilsson AC, Björck IME, et al. Effects of wheat bran extract rich in arabinoxylan oligosaccharides and resistant starch on overnight glucose tolerance and markers of gut fermentation in healthy young adults. European Journal of Nutrition. 2016;**55**(4):1661-1670. DOI: 10.1007/s00394-015-0985-z

[80] Saini A, Panwar D, Panesar PS. Bioactive compounds from cereal and pulse processing. Austin Journal of Nutrition & Metabolism. 2019;**6**(2):1-7

[81] Nguyen NK, Deehan EC, Zhang Z, Jin M, Baskota N, Perez-Muñoz ME, et al. Gut microbiota modulation with long-chain corn bran arabinoxylan in adults with overweight and obesity is linked to an individualized temporal increase in fecal propionate. Microbiome. 2020;**8**(1):1-21. DOI: 10.1186/s40168-020-00887-w

[82] Park HY, Lee KW, Choi HD. Rice bran constituents: Immunomodulatory and therapeutic activities. Food & Function. 2017;**8**(3):935-943. DOI: 10.1039/c6fo01763k

[83] Cian RE, Garzón AG, Martínez-Augustin O, Botto CC, Drago SR. Antithrombotic activity of brewers' spent grain peptides and their effects on blood coagulation pathways. Plant Foods for Human Nutrition. 2018;**73**(3):241-246. DOI: 10.1007/s11130-018-0682-1

[84] Sentís-Moré P, Ortega-Olivé N, Egea-Vilches C, Romero-Fabregat MP. Screening of protein hydrolysates from malt rootlets and their relationship with technological properties. LWT. 2023;**182**:1-9. DOI: 10.1016/j.lwt.2023.114864

[85] Zielke C, Teixeira C, Ding H, Cui S, Nyman M, Nilsson L. Analysis of β-glucan molar mass from barley malt and brewer's spent grain with asymmetric flow field-flow fractionation (AF4) and their association to proteins. Carbohydrate Polymers. 2017;**157**:541-549. DOI: 10.1016/j.carbpol.2016.10.045

Chapter 4

Exploring the Impacts of Climate Change on the Nutritional Properties and Food Security of Various Cereal Grains

Maha Khalfalla and Zoltán Győri

Abstract

Climate change substantially influences agriculture, affecting food security and agricultural production. To address the current concerns, it is essential to address climate-smart agricultural methods, such as crop rotation, integrated pest control and enhanced nitrogen fertilisation techniques, to assist farmers in adjusting to a shifting climate. Furthermore, an ongoing review is being conducted to investigate the potential effects of climate change mitigation and the contribution of agriculture to reducing greenhouse gas emissions abroad. This investigation encompasses various aspects such as agricultural practice and crop varieties, particularly crop relocation, soil nutrient management and innovative nitrogen fertiliser techniques. Restricting the discourse to the crop and N fertiliser selection options and the implementation of various strategies, such as identifying the most resilient crop for climatic fluctuations, implementing a crop relocation system as conventional and modern agricultural practices, minimising the reliance on pesticides and enhancing the nutritional qualities of better cultivars, in addition to the grain drying process and storage, may influence the nutritional composition of cereal grains. All the above adaptation mechanisms depend on the local context, area or country. Ecologically, low-impact solutions that modernise agriculture include biodiversity-based and climate-smart farming. These initiatives aim to effectively enhance agricultural incomes and production while addressing the interrelated challenges of climate change and food security.

Keywords: agriculture practice, grains, buckwheat, crop rotation, proso millet, nutritional contents

1. Introduction

Fluctuations in climate patterns, including factors such as temperature, rainfall and the occurrence of frosts, can either enhance or deteriorate agricultural conditions throughout different regions of the globe. Increased temperatures and elevated CO_2 levels may enhance crop yields in some areas or vice versa [1]. Nevertheless, changes in precipitation patterns, severe weather occurrences and decreased water availability might lead to a decline in agricultural production [2]. The yearly emissions from the global food system, mostly attributed to methane and nitrous oxide emissions from agriculture, account for around 21–37% of total emissions [3, 4]. To ensure food security and sustainable production amidst significant environmental changes, farmers and the agricultural sector may use many ways to adjust to these emerging circumstances [1]. Research suggests that a temperature increase of 2.0°C might result in a 20–40% decrease in cereal grain output due to the impact of climate change, particularly in Asia and Africa [5]. The growth of grains such as wheat, maize and rice is very susceptible to temperature variations [6]. Several management strategies are suggested to guarantee food security and sustainable production in the face of climate change. These include the development of innovative crop varieties, the implementation of efficient cropping systems and the optimisation of nutrient management [5]. Developing climate-resilient crops is a commonly employed approach to breed and grow crop varieties that can withstand the negative impacts of climate change, such as flooding, drought and extreme heat, which can help farmers adapt to changing environmental conditions [7, 8]. Implementing diversified cropping systems and crop rotation techniques in agriculture contributes to ensuring food security and advancing sustainable production in response to the challenges posed by climate change [9, 10]. Prior European study investigated the occurrence of pest insects in different agricultural environments by the use of a multiple-choice methodology [11]. The study evaluates the suitability of crops, the impact of weed species and the effectiveness of crop rotation as a management strategy [12]. By using agricultural diversification practices such as implementing varied crop rotations, utilising cover crops and practising intercropping, it is possible to enhance ecosystem services, minimise environmental damage and maintain crop yields while safeguarding quality [13]. The simulation results demonstrate that the application of nitrogen fertilisation, the timing of planting and the provision of irrigation substantially influence the production of sorghum and millet. These findings indicate that implementing appropriate crop management strategies could effectively reduce the risks associated with climate change [14]. Sorghum, millets and pseudo-cereals are crucial grains for establishing resilient and sustainable food systems in the context of climate change. These grains possess the ability to withstand harsh weather conditions and are highly suitable for dry and semi-arid agroecosystems [15, 16]. Previous research has shown that a crop rotation consisting of wheat, maize and millet and fallow periods during a 4-year cycle in Colorado yields better results than traditional methods to enhance food security and build resistance to climate change [17]. Another investigation performed in Canada discovered that cultivating forage sorghum within a traditional rotation of wheat and grain sorghum and fallow land resulted in enhanced productivity and profitability of the cropping system [18]. Research done in Western Ethiopia revealed that the practice of rotating sorghum and finger millet crops was widespread in the study region [19]. In South Africa, sorghum is often irrigated in rotation with other crops, exacerbating nutrient and water depletion [20]. In Europe, Hungary has optimal conditions for cultivating early and intermediate-maturity sorghum hybrid. Sorghum production is advantageous for crop rotation [21]. The European Union's Green Deal sought to achieve climate neutrality in Europe by 2050. This will be accomplished via the promotion of sustainable food systems, the advancement of organic farming practices and the facilitation of market entry for alternative crops. The European Union's Green Deal aims to achieve climate neutrality in Europe by 2050. Promoting sustainable food systems, advancing organic farming methods, and facilitating market entry for alternative crops will help accomplish this goal, besides utilizing conventional and regionally tailored cultivars of crops [22]. In southern

Hungary, implementing crop rotation systems has the ability to enhance soil health, fertility and nutrient balance, hence favourably impacting the nutritional quality of the crops [23, 24]. Sorghum and millets offer potential for improving food nutritional value, health and sustainable food systems. Biofortification strategies and unique protein functionality are essential [25]. By carefully choosing sorghum and millet types for a rotation system, farmers may establish a more sustainable and productive agricultural system that is specifically suited to their location and market circumstances [26]. The studies emphasised the potential for enhancing productivity and profitability via the significance of grain quality and planting strategies in the choice of crop varieties [27]. Grain storage ensures food security, market stability and profitability. Government agencies store grains to uphold quality standards and minimise losses [3].

The objective of the book chapter was to evaluate the impact of agricultural practices such as grain drying and storage on the resilience of proso millet, sorghum, and buckwheat cereal grains in the face of climate change. Additionally, the chapter examines how these practices affect these grains' nutritional and quality aspects and their potential contribution to high-quality production for human food security.

The scientists and breeders have carefully examined the potential implications of multiple factors on agricultural practices, which may significantly affect production and food security. These considerations are further discussed below.

2. The implications of climate change on agriculture

Climate stress may harm cereal grains' nutritional value, reducing nutrient absorption and assimilation. The repercussions vary according to the crop's classification, the stress level and the prevailing environmental circumstances. Therefore, it is crucial to consider the impact of climatic stress on grain quality while selecting crops and implementing farming practices [28]. The biosecurity toolkit developed by the Food and Agriculture Organisation (FAO) provided strategic guidance for safeguarding food security, commerce and agricultural practices. It specifically addresses global challenges such as environmental and biodiversity difficulties [29].

The diversity of the varieties in agricultural exchange programmes may provide useful insights into the abundance of crop diversity region. It also highlights the need to conserve and use these varieties for the purpose of sustainable agriculture [29–31].

Climate change is significantly impacting Europe's agriculture, affecting crop and livestock production and posing challenges for farmers in adapting to changing conditions, which can present several negative impacts, as are detailed in the following points:

Climate change has prompted significant adaptations in US agriculture, with farmers employing strategies to mitigate crop yields and losses [32]. China's climate change adaptation techniques for maize output are being studied to enhance understanding of the challenges of changing climatic patterns in agriculture [33]. France is known for its rich history of cultivation diversity in Western Europe. Based on the RCP8.5 warming scenario and assuming no changes in growing regions or technology, our model ensemble indicates that winter wheat output would fall by 21.0%, winter barley yield will decline by 17.3% and spring barley yield will decline by 33.6% by the end of the century [34]. Furthermore, climate change will increase the susceptibility of the EU's agri-food business to drought in non-EU nations, with over 44% of the EU's agricultural imports projected to be at high risk of drought in the coming years [35]. China's evidence earlier predicted climate change impacts global agriculture and food production. However, the connection between climate change and national food security must be adequately understood [36]. A study from Africa examined the difficulties and potential advantages of sorghum growing techniques in the humid agroecology of Western Ethiopia [19].

2.1 Alteration and insertion of adaptive crops for climate change

The FAO/OECD Workshop focused on addressing the problems posed by climate change in different agro-ecological and socio-economic settings and building resilience for adaptation in the agriculture sector. The workshop explored resilience from several angles, including biophysical, economic and social dimensions and spanning different levels of analysis, from individual farms to the global scale [37]. A European exploration study has proposed many techniques that might assist farmers and policymakers in formulating suitable adaptation strategies to tackle the problems presented by climate change in the agricultural sector [38, 39]. These strategies are of the utmost importance, such as:

2.1.1 Climate resilience

Employing climate resilient techniques to improve the capacity of agricultural systems to forecast, prepare for, adjust to, withstand and recuperate from the effects of climate change and severe weather events [40].

2.1.2 Climate-resilient practices

The collection of climate-resilient practices may help developers and other stakeholders, and the strategies can be customised to address the individual hazards, levels of exposure and vulnerabilities identified throughout the climate risk assessment process [41]. Policymakers have discovered the optimal crops that can acclimatise to climatic changes and are devising efficient measures to bolster the agricultural industry in response to a shifting climate [42].

2.1.3 Crop switching and relocation

The strategic shifting and relocation of profitable crops for agricultural purposes to adapt to climate change is a crucial component of climate change resistance programmes [26, 43]. Implementing crop switching may mitigate agricultural losses caused by climate change, while relocation can facilitate the identification of appropriate crop adaptation prospects for farmers and policymakers [9]. Implementing a diverse crop rotation may be a very successful approach to enhancing the availability of soil nutrients and preserving soil health in no-till systems. Conversely, windrow-burning waste could harm soil fertility [43], as was explained by **Figure 1**.

3. Incorporating *P. millet*, *Sorghum bicolor*, and common buckwheat into crop rotation strategies

P. millet and *S. bicolor* are scientifically known as the kind of grains that are notable for their capacity to thrive in dry and hot environments, making them valuable for ensuring food security in desert locations, given the anticipated pattern of increasing drought intensity in the Mediterranean region [22, 44]. Buckwheat is a rapidly growing

Figure 1.
Switching and relocation system in the agriculture.

crop that is well-suited for various crop rotation systems, and its nutritional needs should be considered when designing such systems [45]. *P. millet* has been shown to enhance overall production and may serve as a substitute for summer fallow in winter wheat-fallow rotations within crop rotation systems [44]. Studies have shown that *S. bicolor* is a valuable crop in crop rotation systems since it is more likely to achieve the highest yield in certain rotations [14]. Which can provide several advantages, including in the following section details.

4. Implication of *P. millet, S. bicolor* and common buckwheat rotation system on soil fertility

Crop rotation, via the cultivation of several crops in a particular area, may mitigate the depletion of specific nutrients in the soil, resulting in a more equitable and sustainable absorption of nutrients [46].

Crop rotation, which involves the cultivation of *P. millet*, *S. bicolor* and buckwheat, effectively disrupts patterns of nutrient depletion or surplus by modifying the need for certain nutrients. In addition, varied crop rotations enhance microbial diversity, which in turn improves soil structure, nutrient accessibility and breakdown of organic waste [43, 47, 48]. In addition, implementing crop rotation systems using *P. millet*, *S. bicolor* and buckwheat has boosted soil fertility by stimulating microbial populations, improved soil aggregation and facilitated water infiltration, aeration and root development. Consequently, this results in enhanced assimilation of plant nutrients and heightened agricultural output [22, 49]. Buckwheat, classified as a legume, can capture atmospheric nitrogen via the action of nitrogen-fixing bacteria, resulting in the enhancement of soil fertility with this vital nutrient [37]. The wide fibrous root structure of *S. bicolor*, capable of growing to great depths, improves soil aggregation and water penetration, facilitating nutrient absorption and enhancing soil fertility [50, 51]. *P. millet*'s shallow root system is advantageous for nutrient absorption and soil structure enhancement, particularly in less fertile soils [52, 53]. Research conducted by Frontier examined the performance of sorghum hybrids with enhanced terminal senescence under optimal environmental circumstances, resulting

Figure 2.
The role of root structure in facilitating nutrient absorption, letters refer to (a) Sorghum root structure in the soil (b) Sorghum hybrids under favourable environmental conditions (left) and sorghum hybrids grown under resource-poor conditions (right). Source: Copyright © 2022 Ostmeyer, Bahuguna, Kirkham, Bean and Jagadish. This open-access article is distributed under the terms of the Creative Commons Attribution License (CC BY). The use, distribution or reproduction in other forums is permitted, provided the original author(s) and the copyright owner(s) are credited and that the original publication in this journal is cited, in accordance with accepted academic practice. No use, distribution or reproduction is permitted which does not comply with these terms. Correspondence: S. V. Krishna Jagadish, kjagadish@ksu.edu.

in higher nitrogen transfer from leaves to improve both production and grain quality (left), as compared to stay-green sorghum hybrids cultivated in resource-limited situations [53]. Sorghum hybrids that possess rapid nitrogen transfer and exhibit accelerated senescence in less challenging settings may not need a complex root structure [53], as displayed in **Figure 2** (left).

Buckwheat has the capacity to improve soil fertility [54]. In addition, the root exudates have the capacity to mobilise nutrients, particularly phosphorus, by solubilising them, making them more available [55]. Furthermore, the high efficacy of these crops in reducing wireworm populations and promoting soil health is noteworthy [54, 55].

Soil composition, pH levels, and nutrient levels influence cereal grain nutritional composition. Additionally, breeding techniques and agricultural methods significantly enhance grain quality under different environmental conditions, impacting productivity, food security, and human consumption [56]. **Table 1** shows that crop tissues can absorb essential plant nutrients.

5. Pesticide and residue utilisation concepts

The excessive use of pesticides can cause environmental pollution and pose risks to human health. The Food and Agriculture Organisation (FAO) of the World Health Organisation (WHO) establishes maximum residue limits and oversees the use of pesticides in food production to mitigate health and environmental risks [57]. Pesticide usage in food production should be monitored to reduce health and environmental concerns [11]. Research was conducted to analyse the impact of pesticide residues on food safety and the environment to protect consumers' interests and maintain the food supply's stability [58].

Elements	Chemical symbol	Chemical forms uptake by crop
Primary nutrients		
Nitrogen	N	NO_3^-, NH_4^+
Potassium	K	K
Phosphorus	P	HPO_4^-, HPO_4^{2-}, PO_4^{3-}
Secondary nutrients		
Sulphur	S	SO_4^{2-}
Calcium	Ca	Ca^{2+}
Magnesium	Mg	Mg^{2+}
Micronutrients		
Copper	Cu	Cu^+, Cu^{2+}
Iron	Fe	Fe^{3+}
Zinc	Zn	Zn^{2+}
Manganese	Mn	Mn^{2+}
Boron	B	BO_3^{3-}
Molybdenum	Mo	MoO_4^{2-}
Nickel	Ni	Ni^{2+}
Chlorine	Cl	Cl^-

Source: [31].

Table 1.
The elemental nutrients which crops can obtain from the soil source.

Some foods may retain pesticide residues after washing or peeling. Buying organic fruit and growing veggies using Integrated Pest Management (IPM) are the best strategies to limit pesticide exposure in food [58].

5.1 Pesticide utilisation in case of *P. millet, S. bicolor* and common buckwheat

The dynamic crop simulation model focuses on optimising the dynamics of pesticide residues in plant-environment systems, with a particular emphasis on *P. millet*. Drought and heat are excellent methods for minimising residues in *P. millet* crops, making them robust to environmental conditions [59].

Several insecticidal measures were used to control *Melanaphis sacchari*, which may provide useful impacts on pesticide usage in sorghum cultivation. These strategies can help create efficient approaches to managing the aphid and minimise its detrimental impact on sorghum crops [60, 61].

Based on a comparative investigation, buckwheat showed resistance to weed pressure and insect pest damage, making it a good alternative plant for conducting semi-field pesticide toxicity evaluations, such as impatiens flowers [62, 63].

6. Organic and conventional fertiliser concepts in case of *P. millet, S. bicolor* and common buckwheat

Optimised organic manures, chemical fertilisers and bio-priming can enhance soil physicochemical characteristics and *P. millet* yield production [64, 65]. It was proposed that using integrated nutrition sources and seed priming techniques may enhance the development, production and overall quality of nutri-cereal *P. millet* [66]. The combined use of organic amendments and chemical fertilisers has been shown to enhance *S. bicolor*'s growth, productivity and nutritional value. Organic additions are a viable substitute for inorganic fertilisers in improving crop yield and soil quality [65]. Moreover, using organic supplements has enhanced productivity and superior quality of *S. bicolor* in semi-arid tropical locations, exceeding the advantages of mineral fertilisers. This intriguing method can improve soil health and boost crop yield [67].

Favourable results were carried out to evaluate the impact of different farming techniques (organic and conventional) on the nutritional properties of buckwheat varieties grown in Poland [68, 69].

7. Concepts of nitrogen fertiliser impact on nutrient absorption

Crop rotation facilitates nitrogen cycling, hence enhancing nutrient uptake by crops [69]. However, agricultural land use leads to soil nutrient extraction, affecting yield; insufficient replenishment can lead to low nutrient availability. Solutions include fertilisers and crop breeding to overcome this issue [70]. Crop nutrient improvement may be accomplished by diverse breeding techniques and genetic variation, resulting in enhanced productivity, disease resistance, and adaptability to different climates [71].

Nitrogen fertiliser is a crucial component in contemporary agriculture since it has been scientifically shown to enhance plant growth and production [72]. Despite worries about its environmental consequences, including pollution and carbon emissions, and the need for more sustainable alternatives, the prevailing agreement is that, until superior choices become accessible, farmers necessitate unrestricted availability of nitrogen fertiliser to guarantee food security [53]. Nitrogen fertilisers have played a crucial role in contemporary agriculture, facilitating fast development and enhancing productivity in farming [73]. Nevertheless, there are continuous endeavours to discover alternate and more sustainable methods, such as nano-enabled fertilisers, that may enhance nutrient utilisation efficiency and minimise environmental repercussions in contrast to conventional synthetic fertilisers [74]. Although there are difficulties and worries, the need for nitrogen fertiliser in contemporary agriculture is generally acknowledged, and continuous research and innovation strive to tackle its ecological consequences while guaranteeing an uninterrupted food supply [75]. Phenotyping technologies play a crucial role in breeding strategies for sustainable agriculture. They are used to evaluate the capacity of cover crops to withstand changes in the environment and identify genetic ways for enhancing root properties. One such strategy is using buckwheat as a cover crop [76].

The nutritional needs of various crop kinds are determined by their intended use and might vary significantly depending on the adaptability of the crop. Comprehending these essential nutrient needs is vital for efficient nutrient control and for enhancing the ideal development and productivity of various crops [77].

7.1 The employed technologies for efficient N fertiliser utilisation

Modern phenotyping technology and innovative breeding techniques in cover crops have the potential to increase these functions. These tactics focus on enhancing nitrogen fixation, nutrient absorption and stress resistance, among other characteristics, to maximise root growth and performance in agricultural contexts. By incorporating root features into breeding programmes, we want to increase the ability of cover crops to provide ecosystem services, which will eventually contribute to sustainable agriculture and boost soil health [77, 78]. For example, **Figure 3** illustrates: Nano N fertilisers may be administered to crops via three different methods: by being enclosed inside nanomaterials, by being covered with a thin protective polymer coating or by being supplied as particles or emulsions of nanoscale dimensions [79].

Rotating N_2-fixing grain legumes with cereals may provide advantages such as enhanced seed productivity and quality, augmented family monetary revenue and decreased dependence on artificial nitrogen fertilisers [47]. Developing nitrogen fixation in cereals has been a persistent research obstacle; however, advancements have been achieved in the identification of nitrogen-fixing bacteria that have the potential to serve as crop inoculants. Scientists are investigating methods to genetically modify cereal crops to independently produce nitrogen, potentially decreasing reliance on synthetic fertilisers [78].

7.1.1 Optinyte technology

Optinyte technology, developed by Corteva Agriscience, is a nitrogen stabiliser that safeguards nitrogen fertiliser and minimises environmental loss by impeding the nitrification process and limiting the conversion of ammonium to nitrate. It is used in N-Serve and Instinct NXTGEN nitrogen stabilisers to guarantee that the nutrient remains in the root zone throughout crucial periods of growth development [80–82].

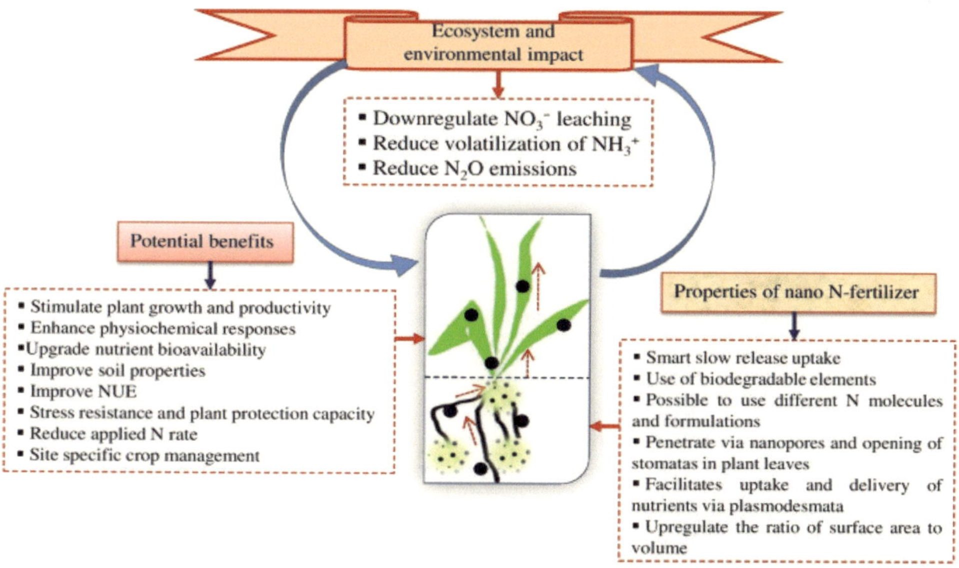

Figure 3.
Advantages of employing the modern technology of nitrogen fertiliser [79]. Source: Springer Nature. Date: Oct 12, 2023. Copyright © 2023, The Author(s) Creative Commons. This open-access article is distributed under the terms of the Creative Commons CC BY license, which permits unrestricted use, distribution, and reproduction in any medium, provided the original work is properly cited. You are not required to obtain permission to reuse this article. CCo applies for supplementary material related to this article, and attribution is not required.

Figure 4.
Mechanism of the Optinyte technology for efficient nitrogen fertiliser utilisation [81]. Source: https://www.thedailyscoop.com/news/retail-industry/two-takeaways-nitrogen-stabilizer-research.

The Optinyte technology has several advantages in agriculture concerning environmental sustainability and agricultural productivity [81, 82]. Several notable advantages are demonstrated in **Figure 4**.

7.1.2 Soilgenic Technologies

Soilgenic Technologies, LLC, has developed a pioneering method that enhances nitrogen fertiliser efficiency and decreases emissions. Additional technologies include the use of slow-release fertilisers, precision agriculture, and cover crops [82].

Slow-release fertilisers gradually release nitrogen over an extended duration, hence minimising the chances of leaching and volatilisation. Precision agriculture uses data and technology to enhance the efficiency of fertiliser application by optimising rates and timing, thereby minimising wastage and enhancing crop absorption. Cover crops may effectively scavenge residual nitrogen in the soil and mitigate leaching [80, 83, 84].

8. Implication of N fertiliser on *P. millet, S. bicolor* and common buckwheat

8.1 Nutrient management

Multiple literature sources have examined the effects of different degrees of nitrogen fertilisation on *S. bicolor* yield and quality characteristics [85, 86]. These studies emphasised the utilisation of nitrogen fertilisation in sorghum production and its ability to enhance yield and quality features [86]. Nevertheless, it is crucial to consider nitrogen fertiliser's environmental ramifications and long-term viability [86, 87].

The 60 kg/ha^{-1} rate of N fertiliser was recommended for *S. bicolor* crops after the rotation of cowpea and soybeans [87]. The physicochemical qualities of *S. bicolor* are influenced by elevated amounts of nitrogen fertiliser, which vary depending on the application stage and the sorghum's specific variety. These effects have a significant impact on the main constituent of *S. bicolor* grains [88–90].

Studies on *P. millet* indicated that a lack of nitrogen has a detrimental effect on grain production, decreasing nitrogen utilisation efficiency and increasing the carbon/nitrogen ratio by lowering the overall nitrogen content and the levels of soluble proteins and sugars [52]. Nitrogen fertilisation rates significantly influenced the development and yield of *P. millet*. In the past, the production and consumption of small millets declined. Nevertheless, they have a crucial function in guaranteeing both the availability of food and a stable source of income, particularly in locations vulnerable to climate change and with limited resources [10, 91]. In contrast, now benefiting from superior morphological and molecular traits of *P. millet* crop as an emergency crop for human consumption is a target in Mediterranean countries [92–94].

The maximum yield was achieved with the highest nitrogen fertilisation rate, highlighting its importance for promoting sustainable agriculture and ensuring food security in Mediterranean areas [94]. Another Mediterranean research demonstrated a distinct impact of nitrogen fertilisation rates on *P. millet* morphology, productivity and phenology. Utilising nitrogen impacted protein content and all other variables, except for the weight of 1000 seeds [94].

Common buckwheat (*Fagopyrum esculentum* M.) is a pseudocereal that plays a crucial role as a catch crop in crop rotation systems [76]. Buckwheat, a crop grown by traditional farmers, has the potential to support smallholder farmers and rural communities [3]. Buckwheat is vital as catch crops can enhance nutrient cycling in cropping systems and reduce nitrogen leaching [48]. The choice of catch crop species and mixture can impact nutrient uptake and carry-over potential, and the growth conditions can also affect nutrient conservation and losses [49, 95]. Legumes can undergo nitrogen fixation by forming a mutually beneficial relationship with rhizobia, while cereals often rely on synthetic nitrogen fertilisers [95].

The utilisation of buckwheat in cover cropping and sorghum grains presents both opportunities and challenges in sustainable agriculture systems in Southern Australia, specifically in Mediterranean climatic conditions. Further evaluation and identification of appropriate cover crop species for the region are necessary [47, 96].

9. Implication of variety diversity role on nutrient management

Various crop varieties have distinct capabilities for absorbing and utilising nutrients, which might influence the effectiveness of nutrient management techniques. Legume crops may effectively mitigate the excessive reliance on chemical fertilisers, particularly nitrogenous fertilisers, in cereal-based cropping systems [97, 98]. The sustainable preservation of genetic resources is essential for enhancing the resilience of the ecosystem, boosting food production and ensuring nutritional security. That requires extensive initiatives to gather and save cultivated varieties, landraces and wild accessions.

Each crop has a distinct nutritional need for achieving optimum development and exhibits different nitrogen, phosphorus and potassium requirements [98], as shown in **Table 1**.

Grain crops such as wheat, barley, oats, rye, and triticale possess unique nutritional demands, with various cultivars within the same crop category sometimes exhibiting diverse requirements, varying from millets and pseudocereal requirements [77].

9.1 Implication of *P. millet, S. bicolor* and common buckwheat variety diversity on nutrient management

S. bicolor germplasm exhibits high genetic diversity, offering potential for cultivar development, breeding strategies and conservation efforts. Genetic regions under selection may contain genes for improved production and adaptation [99, 100]. The discussed factors can contribute to understanding the plasticity and stability of grain sorghum varieties, which can be valuable for developing new cultivars and improving sorghum production in different climatic conditions [100]. The genetic diversity and heritability of sorghum traits provide insights into breeding and conservation efforts and contribute to the knowledge of sorghum genetic resources in the region [101, 102].

Prior investigation indicated that the growing site does not significantly influence the features of *P. millet*. However, the selection of a certain variety may impact its nutritional and functional qualities, which can benefit farmers and researchers [52, 103, 104]. The *P. millet* exhibits a wide range of nutritional characteristics, including protein content (9.5–17%) and vital vitamins and minerals such as vitamin B, iron, calcium, potassium, zinc and magnesium. The farmers and breeders were encouraged to improve the cultivation and development of *P. millet* varieties with specified nutritional needs [105–107].

Previous findings focused on the physical diversity of buckwheat landraces that originated from Northeast India. The research included collecting buckwheat landraces from various elevations, ranging from 103 to 2971 m. The results indicated significant genetic variation within the germplasm of both common buckwheat (*Fagopyrum esculentum* M.) and Tartary buckwheat (*Fagopyrum tataricum*) [97]. In addition, researchers from Europe and Asia used simple sequence repeats (SSR) markers to examine the genetic diversity of several buckwheat species, such as *Fagopyrum esculentum* and *Fagopyrum tataricum*. For the examination of plant genetic diversity, the creation of genetic maps, gene mapping, and cloning are essential due to their locus specificity and co-dominant inheritance, facilitating precise identification and manipulation of target genes [108, 109]. Diverse crop varieties enhance farmers' productivity and sustainability by adapting to changing environmental conditions, as seen in higher yielding common buckwheat varieties with determinate growth habits [110].

9.2 Distinction between the varieties through morphological properties

The variety of the crops is a significant factor to consider when evaluating the multi-element composition of cereal and legume crops [111]. The distinction between cereal grains and a variety may be made by examining their morphological qualities, including grain size, shape, colour and texture [112]. Furthermore, considering the presence of endosperm tissue, kernel hardness and other physical characteristics could help identify cereal grain varieties. The plant's morphological traits are determined by its genetic composition and environmental variables [95, 109].

Geographical origin, variety, harvest season and their interactions are just a few variables that may impact the authenticity of grain and legume crops. These variables may influence the elemental composition of crops, and their impacts can differ based on the particular crop and environment. Gaining comprehension of these aspects and their interplay may aid in evaluating the quality and verification of these agricultural products [95, 109].

10. Comparing the essential nutritional composition of *P. millet, S. bicolor* and common buckwheat

The nutritional makeup of *P. millet*, *S. bicolor* and buckwheat may vary depending on the specific type. However, all three are nutrient-rich grains with a well-rounded combination of macronutrients, vitamins, minerals, and antioxidants. As such, they are beneficial complements to a balanced diet.

Table 2 shows the ranges of the essential nutritional contents in *P. millet*, *S. bicolor* and buckwheat in the whole grains.

11. Implication of handing and storge process on nutritional properties and grain security

The process of drying and storing cereal grains is essential for preserving their quality and preventing any deterioration. The drying process minimises losses caused by factors such as time, infestations or storage and also prolongs the duration during which the crop may be used. The following are key factors to consider while drying and storing cereal grains [96, 113].

Dryness: The dehydration procedure eliminates moisture from cereal grains, guaranteeing optimal safety and minimising agricultural losses. Grain dryers use heated air to provide optimal drying of grains, avoiding the potential for decay and the development of mould or aflatoxins. The drying process must reduce the moisture content of the grain to a range of 10–15% to ensure safe storage [114].

Storage conditions: To store cereal grains safely and for an extended period, it is essential to maintain proper storage conditions. The warehouse should be adequately ventilated and have a moisture level below 50%. Temperature and moisture changes during storage can affect the quality of cereal grains [96, 113].

Nutrient	Proso millet	Sorghum	Buckwheat
Energy (kcal/100 g)	354	339	355
Fat (g/100 g)	3	3.3	7.4
Protein (g/100 g)	12.5	10.4	12
Carbohydrates (g/100 g)	70	75	72.9
Dietary fibre (g/100 g)	2.2	6.3	10
Phosphorus (mg/100 g)	285	287	330
Potassium (mg/100 g)	195	350	450
Magnesium (mg/100 g)	114	165	390
Calcium (mg/100 g)	8	28	110
Iron (mg/100 g)	0.8	4.4	4
Zinc (mg/100 g)	1.5	1.8	0.8

Source: [52, 95, 109].

Table 2.
The essential nutritional content in P. millet, S. bicolor and common buckwheat based on the whole grains.

Packaging: Clean and dry cereals may be kept for 3 years, provided they have undergone appropriate drying treatment; it is important to use specialised agricultural technology to eliminate moisture, which may harm the storage process. Grain dryers use heated air to achieve optimal drying of grains, hence guaranteeing utmost safety and minimising crop wastage [115].

Drying and storage methods: Various methods are used to dehydrate and preserve grain crops for commercial purposes. The choice of the suitable technique relies on variables such as the cereal variety, the accessible resources, and the required level of excellence in the end product [96, 113].

Effective grain storage is crucial for maximising income for farmers and guaranteeing a steady food supply. It is also crucial for maintaining optimal grain quality and minimising storage losses, which are vital to ensuring food security. Hence, managing storage conditions, particularly in preserving quality and minimising losses, is crucial in guaranteeing food security [106].

12. Conclusion

Ancient cereal grains, such as *P. millet*, *S. bicolor* and common buckwheat, have several obstacles in their production, including a dearth of enhanced seed variants, inadequate infrastructure and fluctuating demand in affluent nations. The crop rotation system, including catch crops like buckwheat, *P. millet* and *S. bicolor*, can effectively replace the negative effects of climate change, pesticide residues and nitrogen fertilisers. This strategy is successful in managing nutrients and preserving soil fertility. Additionally, using modern technologies such as Optinyte technology and innovative nitrogen fertiliser methods can further enhance the efficacy of nutrient management—efficient agricultural techniques on a wide scale benefit crop quantity and quality, leading to food security. In addition, developing agricultural practice standards might be further complicated by the limited expertise and resources available to smallholder farmers.

However, drying and storage are also crucial issues since conventional approaches are often demanding in labour and time. *P. millet*, *S. bicolor* and buckwheat are vulnerable to insect infestations and sensitive to mould development, making them less prevalent in rural regions. Consumer acceptability and cultural preferences also impact demand since ancient alternative grains are typically considered inexpensive food options in metropolitan settings. Policy and market restrictions also affect the agricultural techniques for cereal grain production, storage, packaging and consumption of grains with appropriate nutritional compositions. Government policies often prioritise staple crops such as rice and wheat, resulting in restricted allocation of resources towards research and development. The fragmented structure of millet markets and the absence of standardised quality and price generate confusion and distrust among consumers and suppliers.

Conflict of interest

The authors declare they have no conflict of interest.

Author details

Maha Khalfalla[1,2]* and Zoltán Győri[1]

1 Faculty of Agricultural and Food Sciences and Environmental Management, Department of Nutrition and Food Sciences, Institute of Nutrition, University of Debrecen, Debrecen, Hungary

2 Central Laboratory, Ministry of Higher Education and Scientific Research, Khartoum, Sudan

*Address all correspondence to: maha.khalfalla@agr.unideb.hu

References

[1] Höhn JG, Rötter RP. Impact of global warming on European cereal production. CAB Reviews: Perspectives in Agriculture, Veterinary Science, Nutrition and Natural Resources. 2014;**9**:022. DOI: 10.1079/PAVSNNR20149022

[2] Tari I, Laskay G, Takács Z, Poór P. Response of sorghum to abiotic stresses: A review. Journal of Agronomy and Crop Science. 2013;**199**(4):264-274. DOI: 10.1111/jac.12017

[3] Parihar M et al. Reviving the forgotten food network of potential crops to strengthen nutritional and livelihood security in North Western Himalayas. Indian Journal of Agronomy. 2021;**66**:44-59

[4] Sage C. The interconnected challenges for food security from a food regimes perspective: Energy, climate and malconsumption. Journal of Rural Studies. 2013;**29**:71-80. DOI: 10.1016/j.jrurstud.2012.02.005

[5] Fatima Z et al. The fingerprints of climate warming on cereal crops phenology and adaptation options. Scientific Reports. 2020;**10**(1):1-21. DOI: 10.1038/s41598-020-74740-3

[6] Lotze-campen H. Climate change, population growth, and crop production: An overview. In: Crop Adaptation to Climate Change. 2011. pp. 1-11. DOI: 10.1002/9780470960929.ch1

[7] Seres E, Sárvári M. The importance of millet production in regional production, with special emphasis on climate change. Acta Agraria Debreceniensis. 2018;**74**:141-146. DOI: 10.34101/actaagrar/74/1679

[8] Silvia SMG, Alfredo TP, Jorge AR-V, Sócrates LP. Sustainable and technological strategies for basic cereal crops in the face of climate change: A literature review. African Journal of Agricultural Research. 2018;**13**(5):220-227. DOI: 10.5897/ajar2017.12818

[9] Gebre GG, Amekawa Y, Fikadu AA, Rahut DB. Farmers' use of climate change adaptation strategies and their impacts on food security in Kenya. Climate Risk Management. 2023;**40**:100495. DOI: 10.1016/j.crm.2023.100495

[10] Regmi B, Kunwar S, Acharya TD, Gyawali P. Potential of underutilized grain crops in the Western Mountains of Nepal for food and nutrient security. Agriculture. 2023;**13**(7):1360. DOI: 10.3390/agriculture13071360

[11] Leskovac A, Petrović S. Pesticide use and degradation strategies: Food safety, challenges and perspectives. Food. 2023;**12**(14):2709. DOI: 10.3390/foods12142709

[12] Toepfer S, Zellner M, Kuhlmann U. Food and oviposition preferences of *Diabrotica v. virgifera* in multiple-choice crop habitat situations. Entomologia. 2013;**1**:e8. DOI: 10.4081/entomologia.2013.e8

[13] Rodriguez C, Dimitrova Mårtensson LM, Zachrison M, Carlsson G. Sustainability of diversified organic cropping systems—Challenges identified by farmer interviews and multi-criteria assessments. Frontiers in Agronomy. 2021;**3**:698968. DOI: 10.3389/fagro.2021.698968

[14] Araya A et al. Evaluating crop management options for sorghum, pearl millet and peanut to minimize risk under the projected midcentury

climate scenario for different locations in Senegal. Climate Risk Management. 2022;**36**:100436. DOI: 10.1016/j.crm.2022.100436

[15] Teferra TF. Quinoa and other Andean ancient grains: Super grains for the future. Cereal Foods World. 2019;**64**(5):54. DOI: 10.1094/cfw-64-5-0053

[16] Sinkovic L. Underutilized and pseudocereals in the mediterranean diet. Austin Food Sciences. 2016;**1**(5):1021-1022

[17] Magdoff F. Building soils for better. Crops. 1993;**156**(5)

[18] Baumhardt RL, Salinas-Garcia J. Dryland agriculture in Mexico and the U.S. southern great plains. Dryland Agriculture. 2015;**23**:341-364. DOI: 10.2134/agronmonogr23.2ed.c10

[19] Mesfin AH, Girma F. Understanding sorghum farming system and its implication for future research strategies in humid agro-ecologies in Western Ethiopia. Journal of Agriculture and Food Research. 2022;**10**:100456. DOI: 10.1016/j.jafr.2022.100456

[20] Slakie E et al. Agriculture-Environment Series – Sorghum/Millet Systems At-A-Glance Agriculture-Environment Series: Sorghum/Millet and Environment in South Asia & Sub-Saharan Africa EPAR Brief No. 213. Evan School of Public Affairs. Adaptations to Land Constraints 4 E2013. pp. 1-17

[21] Ábrahám ÉB, et al. Effects of Whitening Process on the Quality of Different Grain Sorghum Hybrids. The European Journal of Plant Science and Biotechnology. 2012;**6**(Special Issue 1):90-93

[22] Kakabouki I et al. Introduction of alternative crops in the Mediterranean to satisfy EU Green Deal goals. A review. Agronomy for Sustainable Development. 2021;**41**(6):1-19. DOI: 10.1007/s13593-021-00725-9

[23] Kovács B, Horváth J, Pálmai O, Németh T, Gyori Z. Soil analysis (for plant nutrition) in Hungary: Practice and results. Agrokémia és Talajtan. 2010;**59**(1):125-134. DOI: 10.1556/Agrokem.59.2010.1.15

[24] McKevith B. Nutritional aspects of cereals. Nutrition Bulletin. 2004;**29**(2):111-142. DOI: 10.1111/j.1467-3010.2004.00418.x

[25] Taylor JRN, Schober TJ, Bean SR. Novel food and non-food uses for sorghum and millets. Journal of Cereal Science. 2006;**44**(3):252-271. DOI: 10.1016/j.jcs.2006.06.009

[26] Volsi B, Higashi GE, Bordin I, Telles TS. The diversification of species in crop rotation increases the profitability of grain production systems. Scientific Reports. 2022;**12**(1):1-13. DOI: 10.1038/s41598-022-23718-4

[27] Prasad PVV, Staggenborg SA. Growth and production of sorghum and millets. Soils, Plant Growth and Crop Production. 2011;**2**:1-27

[28] Waraich EA, Ahmad R, Halim A, Aziz T. Alleviation of temperature stress by nutrient management in crop plants: A review. Journal of Soil Science and Plant Nutrition. 2012;**12**(2):221-244. DOI: 10.4067/S0718-95162012000200003

[29] Sufiyan A. The role of biodiversity in food security. International Journal of Scientific Research in Science and Technology. 2022;**1**(1):001-008. DOI: 10.56781/ijsrst.2022.1.1.0021

[30] Bernis-Fonteneau A et al. Farmers' variety naming and crop varietal

diversity of two cereal and three legume species in the Moroccan High Atlas, using DATAR. Sustainability. 2023;**15**(13):371. DOI: 10.3390/su151310411

[31] Geisseler D, Lazicki P, Horwath WR. Plant Tissue Sampling Guidelines. University of California; 2017. p. 3. Available from: https://apps1.cdfa.ca.gov/fertilizerresearch/docs/Plant_Tissue_Sampling.pdf

[32] Raj S, Roodbar S, Brinkley C, Wolfe DW. Food security and climate change: Differences in impacts and adaptation strategies for rural communities in the global south and north. Frontiers in Sustainable Food Systems. 2022;**5**:1-18. DOI: 10.3389/fsufs.2021.691191

[33] Cui X, Xie W. Adapting agriculture to climate change through growing season adjustments: Evidence from corn in China. American Journal of Agricultural Economics. 2022;**104**(1):249-272. DOI: 10.1111/ajae.12227

[34] Gammans M, Mérel P, Ortiz-Bobea A. Negative impacts of climate change on cereal yields: Statistical evidence from France. Environmental Research Letters. 2017;**12**(5):10411. DOI: 10.1088/1748-9326/aa6b0c

[35] Ercin E, Veldkamp TIE, Hunink J. Cross-border climate vulnerabilities of the European Union to drought. Nature Communications. 2021;**12**(1):1-10. DOI: 10.1038/s41467-021-23584-0

[36] Ye L et al. Climate change impact on China food security in 2050. Agronomy for Sustainable Development. 2013;**33**(2):363-374. DOI: 10.1007/s13593-012-0102-0

[37] Meybeck A, Lankoski J, Redfern S, Azzu N, Gitz V, Building Resilience for Adaptation to Climate Change in the Agriculture Sector. 2012

[38] Goulding K, Jarvis S, Whitmore A. Optimizing nutrient management for farm systems. Philosophical Transactions of the Royal Society B: Biological Sciences. 2008;**363**(1491):667-680. DOI: 10.1098/rstb.2007.2177

[39] FAO-Unesco. ISRIC_TechPap20.pdf. Wageningen, 1994. Rome: FAO; 1988. Reprint with corrections. 1990. p. 83, 1988. ISBN 90-6672-057-3

[40] Alvar-Beltrán J, Elbaroudi I, Gialletti A, Heureux A, Neretin L, Soldan R. Climate Resilient Practices: Typology and guiding material for climate risk screening. Rome: FAO; 2021

[41] Abramoff RZ, Ciais P, Zhu P, Hasegawa T, Wakatsuki H, Makowski D. Adaptation strategies strongly reduce the future impacts of climate change on simulated crop yields. Earth's Future. 2023;**11**(4):1-13. DOI: 10.1029/2022EF003190

[42] Jacobs C et al. Climate change adaptation in the agriculture sector in Europe (4/2019). In: EEA Report no. 04/2019. 2019. p. 112. DOI: 10.2800/537176

[43] Passaris N, Flower KC, Ward PR, Cordingley N. Effect of crop rotation diversity and windrow burning of residue on soil chemical composition under long-term no-tillage. Soil and Tillage Research. 2021;**213**:105153. DOI: 10.1016/j.still.2021.105153

[44] Ventura F et al. An assessment of Proso Millet as an alternative summer cereal crop in the Mediterranean Basin. Agronomy. 2022;**609**(12):1-18. DOI: 10.3390/agronomy12030609

[45] Myers RL. How to grow buckwheat. Thomas Jefferson Agricultural Institute. 2002:1-4

[46] Louhar G, Bana RS, Kumar V, Kumar H. Nutrient management technologies of millets for higher productivity and nutritional security. Indian Journal of Agricultural Sciences. 2020;**90**(12):2243-2250

[47] Lengwati DM, Mathews C, Dakora FD. Rotation benefits from N2-fixing grain legumes to cereals: From increases in seed yield and quality to greater household cash-income by a following maize crop. Frontiers in Sustainable Food Systems. 2020;**4**:1-16. DOI: 10.3389/fsufs.2020.00094

[48] Selzer T, Schubert S. Nutrient uptake of catch crops under non-limiting growth conditions. Journal of Plant Nutrition and Soil Science. 2021;**184**(6):709-722. DOI: 10.1002/jpln.202100142

[49] Mohamed AI, Hassan SM, Mahamed SA. Exploring Root Traits for Efficient Early Nutrient Uptake of Sorghum (*Sorghum bicolor* L.) Genotypes Grown under Soil Moisture Stressed and Nutrient Limited Soils: A Root Chamber Experiment. Asian Journal for Advances in Research. 2023;**6**(August):422-436

[50] Gladman N et al. Sorghum root epigenetic landscape during limiting phosphorus conditions. Plant Direct. 2022;**6**(5):1-17. DOI: 10.1002/pld3.393

[51] Narciso JO, Nyström L. The genetic diversity and nutritional quality of proso millet (*Panicum miliaceum*) and its Philippine ecotype, the ancient grain 'kabog millet': A review. Journal of Agriculture and Food Research. 2023;**11**:688. DOI: 10.1016/j.jafr.2023.100499

[52] Gong X et al. Interspecific root interactions and water-use efficiency of intercropped proso millet and mung bean. European Journal of Agronomy. 2020;**115**. DOI: 10.1016/j.eja.2020.126034

[53] Ostmeyer TJ, Bahuguna RN, Kirkham MB, Bean S, Jagadish SVK. Enhancing sorghum yield through efficient use of nitrogen—Challenges and opportunities. Frontiers in Plant Science. 2022;**13**:1-11. DOI: 10.3389/fpls.2022.845443

[54] Hansen V, Meilvang LV, Magid J, Thorup-Kristensen K, Jensen LS. Effect of soil fertility level on growth of cover crop mixtures and residual fertilizing value for spring barley. European Journal of Agronomy. 2023;**145**:126796. DOI: 10.1016/j.eja.2023.126796

[55] Wendling M, Büchi L, Amossé C, Sinaj S, Walter A, Charles R. Influence of root and leaf traits on the uptake of nutrients in cover crops. Plant and Soil. 2016;**409**(1-2):419-434. DOI: 10.1007/s11104-016-2974-2

[56] Alloway BJ. Soil factors associated with zinc deficiency in crops and humans. Environmental Geochemistry and Health. 2009;**31**(5):537-548. DOI: 10.1007/s10653-009-9255-4

[57] Food and Agriculture Organization of the United Nations (FAO), World Health Organization (WHO). The International Code of Conduct on Pesticide Management. World Health Organization; 2014

[58] Gerage JM, Meira APG, da Silva MV. Food and nutrition security: Pesticide residues in food. Nutrire. 2017;**42**(1):1-9. DOI: 10.1186/s41110-016-0028-4

[59] Song MH, Yu JW, Keum YS, Lee JH. Dynamic modeling of pesticide residue in proso millet under multiple application situations. Environmental Pollution.

2023;**334**:121993. DOI: 10.1016/j.envpol.2023.121993

[60] Faris AM, Brewer MJ, Elliott NC. Parasitoids and predators of the invasive aphid *Melanaphis sorghi* found in sorghum and non-crop vegetation of the sorghum agroecosystem. Insects. 2022;**13**(7):606. DOI: 10.3390/insects13070606

[61] Bowling RD et al. Sugarcane aphid (Hemiptera: Aphididae): A new pest on sorghum in North America. Journal of Integrated Pest Management. 2016;**7**(1):1-13. DOI: 10.1093/jipm/pmw011

[62] Pokharel SS, Yu H, Fang W, Parajulee MN, Chen F. Intercropping cover crops for a vital ecosystem service: A review of the biocontrol of insect pests in tea agroecosystems. Plants. 2023;**12**(12):126034. DOI: 10.3390/plants12122361

[63] Gradish AE, Cutler GC, Frewin AJ, Scott-Dupree CD. Comparison of buckwheat, red clover, and purple tansy as potential surrogate plants for use in semi-field pesticide risk assessments with *Bombus impatiens*. PeerJ. 2016;**2016**(7):1-17. DOI: 10.7717/peerj.2228

[64] Gavit HD, Rajemahadik VA, Bahure GK, Jadhav MS, Thorat TN, Kasture MC. Effect of establishment techniques and sowing time on yield and yield attributes of proso millet (*Panicum miliaceum* L.). International Journal of Current Microbiology and Applied Sciences. 2017;**6**(5):1523-1528. DOI: 10.20546/ijcmas.2017.605.166

[65] Amujoyegbe BJ, Opabode JT, Olayinka A. Effect of organic and inorganic fertilizer on yield and chlorophyll content of maize (*Zea mays* L.) and sorghum *Sorghum bicolour* (L.) Moench. African Journal of Biotechnology. 2007;**6**(16):1869-1873. DOI: 10.5897/ajb2007.000-2278

[66] Sumalata, Byadgi, Sahana N, Siddaraju R. Influence of integrated nutrient sources and seed priming on growth seed yield and quality in Nutri-cereal Proso millet. Journal of Pharmacognosy and Phytochemistry. 2020;**9**(2):1074-1078. Available from: www.phytojournal.com

[67] Kusvuran A, Bilgici M, Kusvuran S. The effect of different organic matters on plant growth regulation and nutritional components under salt stress in sweet sorghum [Sorghum bicolor (L.) Moench.] Maydica. 2021;**66**(4):1-9

[68] Žvikas V, Pukelevičienė V, Ivanauskas L, Pukalskas A, Ražukas A, Jakštas V. Variety-based research on the phenolic content in the aerial parts of organically and conventionally grown buckwheat. Food Chemistry. 2016;**213**:660-667. DOI: 10.1016/j.foodchem.2016.07.010

[69] Mohler CL, Johnson SE. Crop Rotation on Organic Farms A Planning Manual. Sustainable Agriculture Research and Education (SARE) Program; 2009

[70] Nieves-Cordones M, Rubio F, Santa-María GE. Editorial: Nutrient use-efficiency in plants: An integrative approach. Frontiers in Plant Science. 2020;**11**:10-12. DOI: 10.3389/fpls.2020.623976

[71] Loskutov IG. Advances in cereal crops breeding. Plants. 2021;**10**(8):1142. DOI: 10.3390/plants10081705

[72] Pasley HR, Cairns JE, Camberato JJ, Vyn TJ. Nitrogen fertilizer is an essential input in modern agriculture as it has been shown to increase plant

growth and yield. Nutrient Cycling in Agroecosystems. 2019;**115**(3):373-389. DOI: 10.1007/s10705-019-10016-1

[73] Chen B, Liu E, Tian Q, Yan C, Zhang Y. Soil nitrogen dynamics and crop residues. A review. Agronomy for Sustainable Development. 2014;**34**(2):429-442. DOI: 10.1007/s13593-014-0207-8

[74] Seleiman MF, Almutairi KF, Alotaibi M, Shami A, Alhammad BA, Battaglia ML. Nano-fertilization as an emerging fertilization technique: Why can modern agriculture benefit from its use? Plants. 2021;**10**(1):1-27. DOI: 10.3390/plants10010002

[75] Kelly RM, Strong WM, Jensen TA, Butler D. Application of probability analysis to assess nitrogen supply to grain crops in Northern Australia. Precision Agriculture. 2004;**5**(2):95-110. DOI: 10.1023/B:PRAG.0000022356.01537.67

[76] Griffiths M et al. Optimisation of root traits to provide enhanced ecosystem services in agricultural systems: A focus on cover crops. Plant, Cell & Environment. 2022;**45**(3):751-770. DOI: 10.1111/pce.14247

[77] Vinod KK. The Need for Nutrient Efficient Crop Varieties. OSF. 8 Jan 2019:1705. osf.io/mb9vu

[78] Rosenblueth M et al. Nitrogen fixation in cereals. Frontiers in Microbiology. 2018;**9**:1-13. DOI: 10.3389/fmicb.2018.01794

[79] Verma KK, Song XP, Degu HD, Guo DJ, Joshi A, Huang HR, et al. Recent advances in nitrogen and nano-nitrogen fertilizers for sustainable crop production: A mini-review. Chemical and Biological Technologies in Agriculture. 2023;**10**(1):1-14. DOI: 10.1186/s40538-023-00488-3

[80] Tâm T, et al. Proceeding of the 20th Nitrogen workshop. Ouvent des Jacobins Conference Centre. 2018;**01**:17-472

[81] Corteva. DF-N-Serve-Instinct-Optinyte-Brochure.pdf2004. DOI: 10.10231B:FRES.0000025287.5265.99

[82] Clark A, Matthews M. Canadian agri-food sustainability: Skilled talent needed to meet food demand and reduce environmental impacts. Information and Communications Technology Council (ICTC). Ottawa, Canada. 2023

[83] Ajayi EO et al. Implementing Green Chemistry Principles for Pollution Control to Achieve Environmental Sustainability—A Review. Vol. 11. London, UK: Intech; 2016. p. 13. Available from: https://www.intechopen.com/books/advanced-biometric-technologies/liveness-detection-in-biometrics

[84] Zhou Y, Huang J, Li Z, Wu Y, Zhang J, Zhang Y. Yield and quality in main and ratoon crops of grain sorghum under different nitrogen rates and planting densities. Frontiers in Plant Science. 2022;**12**:778663. DOI: 10.3389/fpls.2021.778663

[85] Kumar R et al. Nutrient uptake and content in sorghum cultivars (*Sorghum bicolor* L) under summer environment. Indian Journal of Plant Physiology. 2017;**22**(3):309-315. DOI: 10.1007/s40502-017-0306-z

[86] Bartzialis D, Giannoulis KD, Gintsioudis I, Danalatos NG. Assessing the efficiency of different nitrogen fertilization levels on Sorghum yield and quality characteristics. Agriculture. 2023;**13**:1253. DOI: 10.3390/agriculture13061253

[87] Yamoah CF, Clegg MD, Francis CA. Rotation effect on sorghum response

to nitrogen fertilizer under different rainfall and temperature environments. Agriculture, Ecosystems and Environment. 1998;**68**(3):233-243. DOI: 10.1016/S0167-8809(97)00152-7

[88] Sunitha V, Kosnam K. Grain sorghum response with different levels of fertilizers. International Journal of Agricultural Sciences. 2021

[89] Bhutada PO, Jahagirdar JE, Kote GM. Response of different fertilizer level to sweet sorghum cultivars in rainfed environment (*Sorghum bicolor* L. Moench). International Journal of Current Microbiology and Applied Sciences. 2020;**9**(6):645-649. DOI: 10.20546/ijcmas.2020.906.082

[90] Singh A, Kumar M, Shamim M. Importance of minor millets (Nutri Cereals) for nutrition purpose in present scenario. International Journal of Chemical Studies. 2020;**8**(1):3109-3113. DOI: 10.22271/chemi.2020.v8.i1au.9226

[91] Muthamilarasan M, Prasad M. Small millets for enduring food security amidst pandemics. Trends in Plant Science. 2021;**26**(1):33-40. DOI: 10.1016/j.tplants.2020.08.008

[92] Chaudhary S, Singh A, Prasad RK, Kumar A, Kaushal R, Negi PS. A short review on millets: A potential nutricereals. The Pharma Innovation Journal. 2020;**9**(10):123-126. Available from: http://www.thepharmajournal.com

[93] Lágler R et al. Morphological and molecular analysis of common millet (*P. miliaceum*) cultivars compared to an aDNA sample from the 15th century (Hungary). Euphytica. 2005;**146**(1-2):77-85. DOI: 10.1007/s10681-005-5814-7

[94] Palchetti E et al. Effects of nitrogen fertilization and plant density on proso millet (*Panicum miliaceum* L.) growth and yield under mediterranean pedoclimatic conditions. Agriculture. 2023;**13**(9):1657. DOI: 10.3390/agriculture13091657

[95] Guo K, Yang J, Yu N, Luo L, Wang E. Biological nitrogen fixation in cereal crops: Progress, strategies, and perspectives. Plant Communications. 2023;**4**(2):100499. DOI: 10.1016/j.xplc.2022.100499

[96] Kumar S et al. Efficient nutrient management for enhancing crop productivity, quality and nutrient dynamics in lentil (*Lens culinaris* Medik.) in the semi-arid region of northern India. PLoS ONE. 2023;**18**(2):1-25. DOI: 10.1371/journal.pone.0280636

[97] Wuest SE, Peter R, Niklaus PA. Ecological and evolutionary approaches to improving crop variety mixtures. Nature Ecology & Evolution. 2021;**5**(8):1068-1077. DOI: 10.1038/s41559-021-01497-x

[98] Anas M et al. Fate of nitrogen in agriculture and environment: Agronomic, eco-physiological and molecular approaches to improve nitrogen use efficiency. Biological Research. 2020;**53**(1):1-20. DOI: 10.1186/s40659-020-00312-4

[99] Wondimu Z, Dong H, Paterson AH, Worku W, Bantte K. Genetic diversity, population structure, and selection signature in Ethiopian sorghum [*Sorghum bicolor* L. (Moench)] germplasm. G3: Genes, Genomes, Genetics. 2021;**11**(6). DOI: 10.1093/g3journal/jkab087

[100] Antimonova O, Syrkina L, Antimonov A, Kosykh L. Assessment of plasticity and stability of grain sorghum varieties on the basis of their grain yield. BIO Web of Conferences. 2020;**27**:00027. DOI: 10.1051/bioconf/20202700027

[101] Miriti P, Regassa MD, Ojiewo CO, Melesse MB. Farmers' preferences and willingness to pay for traits of sorghum varieties: Informing product development and breeding programs in Tanzania. Journal of Crop Improvement. 2022;**37**(2):1-20. DOI: 10.1080/15427528.2022.2079038

[102] Tyl C, Marti A, Hayek J, Anderson J, Ismail BP. Effect of growing location and variety on nutritional and functional properties of proso millet (*Panicum miliaceum*) grown as a double crop. Cereal Chemistry. 2018;**95**(2):288-301. DOI: 10.1002/cche.10028

[103] Braun EA, Scholz H. Plant Guide—Proso Millet. Vol. 46. United States Department of Agriculture. 2014. pp. 5-7

[104] Hegde SC, Desai SR, Naik BK. Standardization and Nutrient Composition of the Proso Millet Chakli. Vol. 132021. pp. 44-50

[105] Qaisrani SN et al. Variability in millet: Factors influencing its nutritional profile and zootechnical performance in poultry. Journal of Applied Poultry Research. 2019;**28**(2):242-252. DOI: 10.3382/japr/pfy073

[106] Sruthi NU, Rao PS. Effect of processing on storage stability of millet flour: A review. Trends in Food Science and Technology. 2021;**112**:58-74. DOI: 10.1016/j.tifs.2021.03.043

[107] Hou S et al. Genetic diversity of buckwheat cultivars (*Fagopyrum tartaricum Gaertn.*) assessed with SSR markers developed from genome survey sequences. Plant Molecular Biology Reporter. 2016;**34**(1):233-241. DOI: 10.1007/s11105-015-0907-5

[108] Pirzadah TB, Malik B. Pseudocereals as super foods of 21st century: Recent technological interventions. Journal of Agriculture and Food Research. 2020;**2**:100052. DOI: 10.1016/j.jafr.2020.100052

[109] Heuermann D et al. Catch crop mixtures have higher potential for nutrient carry-over than pure stands under changing environments. European Journal of Agronomy. 2022;**136**:126504. DOI: 10.1016/j.eja.2022.126504

[110] Grahić J et al. Genetic relationships and diversity of common buckwheat accessions in Bosnia and Herzegovina. Agronomy. 2022;**12**(11):233-241. DOI: 10.3390/agronomy12112676

[111] Zhao H, Tang J, Yang Q. Effects of geographical origin, variety, harvest season, and their interactions on multi-elements in cereal, tuber, and legume crops for authenticity. Journal of Food Composition and Analysis. 2021;**100**:103900. DOI: 10.1016/j.jfca.2021.103900

[112] Welch RW. Cereal grains, Encyclopedia of Human Nutrition; 2005. DOI: 10.1016/B0-12-226694-3/00050-8

[113] Lewis KL, Boutton T. Building Soil and Food Security in Wheat Production Systems. Office of Graduate and Pro-fessional Studies of Texas A&M University; 2021

[114] Said PP, Pradhan RC. Food grain storage practices-a review. Journal of Grain Processing and Storage. 2014;**1**(1):1-5

[115] Jayas DS, White NDG. Storage and drying of grain in Canada: Low cost approaches. Food Control. 2003;**14**(4):255-261. DOI: 10.1016/S0956-7135(03)00014-8

Chapter 5

Innovation in Nixtamalization by Extrusion Using the Wet Process

Carlos Martín Enríquez-Castro, Brenda Contreras-Jiménez and Eduardo Morales-Sánchez

Abstract

Extrusion wet milling nixtamalization (EWMN) is an innovative process that combines traditional nixtamalization with wet extrusion technology to produce high-quality corn products. As is known, wet extrusion technology is an HTST (high-temperature short-time) process and nixtamalization is an LTLT (low-temperature long-time) process. So, EWMN is the combination of these two technologies. It is used in high moisture, low temperature, low screw velocity and the corn grain is milled at 3 mm size. EWMN is based on mixing corn with water and lime, creating a homogeneous masa that is subjected to an extrusion process using a screw or double screw. The operating parameters, such as humidity, cooking time, production speed, and shear, are critical in this process and must be carefully controlled to obtain the desired product texture and characteristics. After extrusion, the product is dried to reduce humidity to safe and desirable levels for storage. This step is essential to increase the life of the final product. In summary, corn nixtamalization extrusion combines the traditional nixtamalization technique with wet extrusion, resulting in high-quality corn products, better digestibility, and efficiency compared to conventional processes. This innovative approach offers a promising solution for corn-based food production.

Keywords: extrusion, nixtamalization, corn masa, high-temperature short-time, low-temperature long-time

1. Introduction

1.1 Context of nixtamalization

The traditional nixtamalization is a centuries-old technique that involves the cooking of corn in an excess of water (3:1) and lime in a concentration of 1–3% w/w, and times from 30 min to 1 h depending on the hardness of the corn. After cooking, the grain is soaked for times from 12 to 24 h in the alkaline solution used for cooking, and a final rinsing of the cooked corn to remove cooking liquid and the pericarp of grains. The process discards the water used for cooking, which has a pH around 12. The product of nixtamalization, the cooked corn known as "nixtamal" is usually ground to create corn masa, the base for making tortillas [1]. Fresh masa is crafted in small local businesses, as well as in modern, highly efficient facilities in Mexico, and other countries of the American continent [2]. Additionally, fresh corn masa can be dried and ground to produce corn flour, which only needs to hydrate for different preparations [3]. The consumption of nixtamalized corn products also has been extended to other countries in Europe, Africa, and the Asian continent. The interest in nixtamalized products is based on their versatility and also its nutritional value, which includes calcium disponibility and important fiber content. On the other hand, industrial-scale production of nixtamalized corn flour (NCF) reduces the use of water, lime, and steeping time. It involves processing nixtamal, drying it, and subjecting it to multiple milling stages to achieve a fine particle size in the flour, resulting in a product that hydrates easily [4].

In our current context of consuming corn flour products, the traditional nixtamalization process continues to be an important method for producing corn masa, tortillas, and related items. However, industrial processes have evolved to optimize efficiency and product quality, making it more accessible to consumers. Additionally, there is a growing emphasis on sustainable and resource-efficient production methods, aligning with contemporary concerns about the environment and health.

The wet nixtamalization process represents a groundbreaking innovation in the traditional technique of nixtamalization. This modern approach involves cooking corn in a slurry of water and lime, and then using advanced extrusion technology to transform it into a versatile corn masa. This innovative method not only reduces the consumption of water but shortens processing time considerably. The resulting corn masa, rich in flavor and nutritional value, offers a sustainable alternative for producing tortillas and various corn-based products. Moreover, the fine particle size achieved through extrusion ensures that the masa hydrates easily, making it suitable for efficient large-scale production. This groundbreaking approach to nixtamalization opens the door to more resource-efficient and environmentally conscious methods of meeting the demand for corn flour products in our modern world.

2. Evolution of nixtamalization

Nixtamalization has undergone a significant evolution over the years, transitioning from a traditional technique to an innovative technologically driven process [5]. This evolution has led to the development of new methods that enhance efficiency [6], reduce resource consumption [7], and adapt to the dynamic demands of our contemporary world.

Emerging technologies addressing the environmental challenges associated with nixtamalization effluents have been introduced. Ramírez-Jiménez and Castro-Muñoz [8] have identified valuable applications for nejayote, a byproduct traditionally discarded during the washing process. Furthermore, alternative techniques, including rotary reactor with steam pressure, ohmic heating, microwave, and extrusion, which eliminates excessive water use through dry milling for starch gelatinization, have been thoroughly explored. The latter is particularly noteworthy for its ability to rapidly gelatinize starch and generate a significant amount of resistant starch, attributed to the abrupt changes experienced by starch molecules [9].

Table 1 presents various technologies proposed for use in the nixtamalization process. In contrast, the traditional method involves a high quantity of water and an extended processing time, encompassing multiple operations such as cooking, steeping, washing, and grinding. The alternative methods proposed operate at the industrial or semi-industrial level and aim to address ecological issues related to water waste and prolonged processing times, often involving modifications in the energy source.

Some of these methodologies are presently used in processing various foods with a few applied in industrial-level nixtamalization. However, a substantial portion of these alternatives still exists as laboratory prototypes. This is primarily attributed to technical challenges, including complexities in installation and difficulties in scaling up to an industrial or semi-industrial level. Conversely, accessible equipment, such as microwaves, is suitable for the general population, with limitations primarily related to processing capacity or accessibility in rural regions.

The methods described in **Table 1** show a general overview of nixtamalization, and every technique has both advantages and disadvantages. However, of these technologies, extrusion is one of the most promising methods due to its versatility and its existence of commercial equipment in the market, which makes easier its use with different modalities of rate, moisture, particle size, die, temperature control, shear rate, and others.

2.1 Importance of innovation in the extrusion nixtamalization technology

In recent times, a groundbreaking innovation has emerged, known as "wet process extrusion nixtamalization." This modern approach combines the principles of wet traditional nixtamalization with advanced extrusion technology. It involves cooking corn in a slurry of water and lime, and then using extrusion to transform it into a versatile corn masa. This innovation not only reduces water consumption but also significantly cuts down processing time. The resulting masa is not only efficient for large-scale production but also rich in flavor and nutritional value.

This evolution in nixtamalization aligns with our current concerns about sustainability and resource efficiency. The utilization of extrusion technology offers an environmentally conscious alternative while maintaining the authentic flavors and qualities of traditional nixtamalization. This transformation in the nixtamalization process reflects the adaptability and ingenuity of culinary traditions to meet the demands of our contemporary society. As we move forward, it is innovations, such as these that bridge the gap between tradition and innovation, ensuring that the essence of nixtamalization endures while catering to the evolving tastes and needs of our world.

3. Nixtamalization in wet extrusion grinding

3.1 Dry process vs. wet process

Nixtamalization is a traditional Mesoamerican food preparation technique used for the processing of maize (corn) into various products, most notably masa dough for making tortillas and other maize-based dishes. There are two primary methods of nixtamalization: the dry process and the wet process. These methods differ in their approach and the outcomes they yield. The differences between the two are the following:

Technology	Specifications	References
Traditional nixtamalization process	Nixtamalization at home involves cooking whole corn in a pot with a water to corn ratio of 3:1, adding 1–3% lime, and boiling for 30–60 minutes. After, cooking corn called nixtamal is steeping for 12–24 hours, and finally, it is washed and ground to create fresh masa.	[10]
	The industrial process in Mexico is also rooted in the traditional method, utilizing large kettles heated by gas to facilitate a semicontinuous process. This results in the generation of effluents typical of the traditional process. The nixtamalized corn flour industry has managed to reduce resting times to 4 hours. This reduction is attributed to the subsequent grinding of the material, inducing gelatinization through mechanical stress, followed by drying to produce instant corn flour.	[11]
Extrusion for nixtamalization with a single screw	The electrically powered equipment processes corn with lime and moisture, using minimal water for starch gelatinization and producing no effluents. It features a short residence time, a single screw mechanism with automatic feeding, simultaneous material transportation and mixing, all under controlled temperature and velocity. The final die shapes the nixtamalized masa for tortillas. Material transport and cooking occur in a heated chamber. A die induces final gelatinization, producing fresh corn masa. The process takes 1.5 to 7 minutes, with temperatures from 70 to 120°C.	[12]
Steam pressure by rotatory reactor	This method employs gas as a heat source to generate steam through combustion. The equipment includes time control to inject a steam flow, ensuring proper heat distribution in grains, with temperatures ranging from 130 to 300°C. Steam supply can be continuous or intermittent. Nixtamalization begins at 70°C and gradually increases to 90°C, inducing changes in corn grains. Additionally, there is a posttreatment step involving the cooling of nixtamalized material with water at 55°C.	[13]
Ohmic heating	Operating in batch mode with electric power, the equipment ensures temperature control through precise thermocouples and systematic power regulation. The principle involves passing an electric current through the food, rapidly heating a mixture of milling corn with minimal water and lime, resulting in the swift production of fresh masa in a matter of seconds.	[1]
Microwave oven	The kitchen-sized equipment operates in batch mode with limited processing capacity, incorporating controls for power and residence time. It utilizes electric energy to produce radiation for cooking corn grits with water and lime. A milling process follows to obtain fresh corn masa. However, this system lacks temperature control.	[14]
Low-shear transport system	This equipment is inspired by extrusion but departs from conventional screw-based transportation by utilizing rolls in the chamber. This change ensures a gentle mixing process, preventing excessive shear rates that could lead to over gelatinization. The method effectively produces fresh corn masa suitable for tortillas or chips. The raw material consists of ground corn mixed with lime and water, and the process operates at temperatures ranging from 75 to 90°C.	[15]

Table 1.
Diverse innovations in corn grain treatment: A comprehensive exploration of techniques.

3.1.1 Dry process

In the context of food extrusion, the "dry process" can be described as an extrusion method that relies on low water content. Maize kernels (whole or partially de-husked) are heated with a few quantity of water, but not completely dried or toasted. Therefore, we refer to dry process when humidity added is at most 30%. In the extrusion nixtamalization corn processing (ENP), dry milling is employed for size reduction, which necessitates a higher level of mechanical effort compared to the wet milling used in the traditional nixtamalization process (TNP). The elevated temperatures and reduced water content within the extruder expedite the gelatinization of starch granules [16]. Both the thermal impairment suffered by starch granules during the drying process and the mechanical damage resulting from dry milling significantly impact their functional properties. The importance of particle size in corn flour production cannot be overstated as it directly influences the gelatinization process and the ultimate quality of the product.

Dry milling is employed to produce both flour and germ, allowing for the optimal isolation of endosperm. This method exerts an abrasive force while removing the germ and pericarp layers. The resulting corn flour finds application in the production of snacks and extruded foods with semolina and various types of flour being the primary products derived from this process [17]. The isolated endosperm undergoes a size reduction process and is categorized based on particle size (coarse, medium, or fine). The moisture content is kept to a minimum, and the size reduction technique may involve various types of mills, including ball mills, hammer mills, or pin mills [18].

3.1.2 Extrusion in food using a screw

During the milling process, it is crucial to consider the type of force applied, whether it be impact, cutting, or compression. The friction generated within the mill depends on the spacing between volcanic stones and the time allocated to achieve the appropriate particle size distribution. This results in an increase in the temperature of the dough, typically ranging from 55 to 85°C, thereby promoting a secondary gelatinization process [19]. The particle size obtained, which is larger than that achieved in dry milling, allows for starch granules to be broken down or occasionally remain in their native state [20].

In contrast to the traditional nixtamalization process, where whole corn kernels undergo alkaline cooking, extrusion nixtamalization requires grinding and conditioning the corn having low moisture content lime. While this process does facilitate chemical interactions between calcium and corn components, it leads to more extensive starch damage due to the use of dry milling. The extent of starch damage depends on the specific equipment employed and the milling conditions applied as dry milling significantly disrupts the crystalline structure of starch granules.

3.1.3 Food extrusion using double screw

Food extrusion with double screws is a versatile and efficient food processing technique. It combines, cooks, and shapes ingredients simultaneously using twin screws. This process offers control over texture and expands product possibilities, making it ideal for snacks, cereals, and various food products. It is energy-efficient, adaptable for recipe customization, and allows for the addition of nutritional elements, all while ensuring consistent product quality through advanced monitoring systems [21].

3.2 Wet traditional nixtamalization process

The traditional nixtamalization is considered a wet process because it involves boiling maize kernels in an alkaline solution prepared with around three parts of water for each part of maize and using lime or ash in 1–3%. The key differences and characteristics of the wet traditional nixtamalization process include boiling in the alkaline solution, a resting of kernels in the cooking liquor (*Nejayote*) where through steeping time, the outer pericarp of corn grain is softened, and nutrient availability improved, such as calcium and essential amino acids [22]. Functionally, the pericarp of the corn is more easily removed during washing, and it makes it easier to grind into masa dough. All traditional nixtamalization involves a high quantity of water, which decreases the shear stress in grains and starch compared with the dry method. Besides, this wet process is extensively known in Mexico and other countries of America to produce masa with a neutral flavor profile and good machinability for applications to applications like making tortillas, or flours, as is shown in **Figure 1**.

During the traditional nixtamalization process, it is possible to have good-quality masa and tortillas; however, the waste of water is a big problem, as well as the loss of fiber of corn which is thrown away in nejayote. On the other hand, the functional and nutritional changes in corn grain and starch only can be reflected through a long time of cooking and steeping due to the use of whole kernels of maize [23]. The functional characteristics of corn masa obtained by nixtamalization include machinability to form the tortilla, with textural properties such as a high cohesion of corn masa, a medium hardness of corn masa, and low values of adhesiveness.

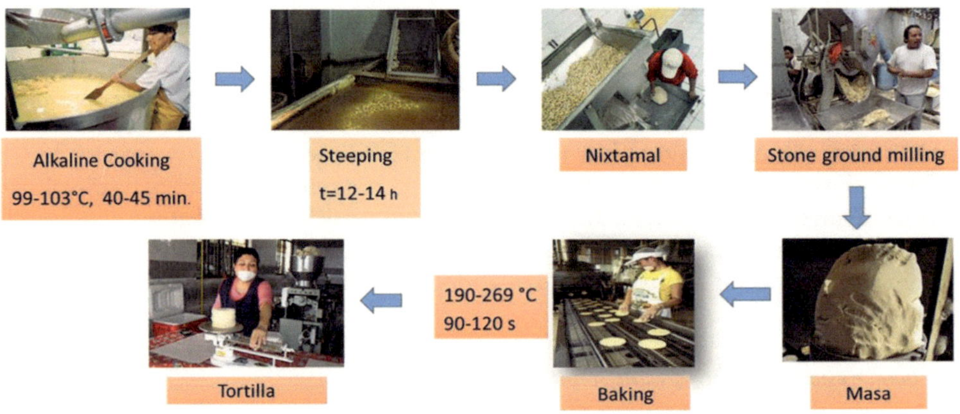

Figure 1.
Traditional nixtamalization process (adapted from [10]).

4. Operating factors and parameters in nixtamalization using extrusion

Extrusion nixtamalization in the past has been considered a process that can implicate an excess of shear stress in starch and other components of corn, which could negatively affect the quality of corn masa in terms of machinability and functionality. However, changes in the operating parameters can modify the quality of corn masa for different nixtamalized products, for example, corn chips or tortillas.

It is well knowing that the advantages of using extrusion for the nixtamalization include unit operations like mixing, transport, cooking, and forming, in addition to not generating polluting effluents. Furthermore, it leads to a reduction in space and energy requirements, minimal water consumption, and the proper mixing of ingredients [24]. On the other hand, trained personnel are required to handle the equipment, the standardization of the process, and the necessary preparation time demand careful attention. Adequate facilities are also needed [25]. Despite these challenges, the environmental and operational benefits make extrusion nixtamalization an appealing option in food production.

The challenge of applying extrusion for the nixtamalization process lies in the accessibility of the equipment, in extruders with different capacities, from home level to semi-industrial and industrial levels, acceptability of the population about the advantages of use of extrusion, and finally to find the better process parameters for production of nixtamalized products.

4.1 Nixtamalized corn masa parameters

The operating factors and parameters in corn masa composition for TNCF and ENCF include corn variety and quality. The type and quality of corn influences the flavor, texture, and nutritional value of the nixtamalized corn flour. Selecting a suitable corn variety with optimal is crucial [26, 27]. The nixtamalization process involves treating corn with an alkaline solution, usually lime (calcium hydroxide). Factors, such as lime concentration, soaking time and cooking temperature play a pivotal role in the flavor and nutritional content of nixtamalized masa [28]. Extrusion involves subjecting a whole flour of corn with a variety of particle sizes mixed with the lime in concentrations from 0.1 to 0.5% w/w and from 30 to 60% of moisture. The masa is mixed, heated, pressured, and transported with the screw and passing through a die, which implies a profile of shear forces on the corn components, mainly starch, that allow to obtain fresh corn masa. Parameters such as extrusion temperature, screw speed, and residence time impact the physicochemical changes in the masa, influencing its texture, functional properties, and flavor.

Functional properties are important parameters from a commercial point of view mainly for the instant nixtamalized corn flour industry, which demands a high-water absorption capacity of flours that increase the yield in terms of conversion from flour to corn masa for making tortillas [29]. The water absorption capacity is also related to the starch damage due to the process, and it significantly affects the textural properties such as hardness and adhesiveness of corn masa, and therefore the tortilla quality. The corn masa should have cohesiveness and low adhesiveness to ensure easier molding [30]. The grinding of nixtamalized corn is a critical step that influences the particle size and distribution in the resulting flour, functional properties, and machinability. A consistent grind ensures uniform hydration and texture in the dough. The grinding of nixtamalized corn is done through wet milling, leading to increased gelatinization and the release of hydrated granules. During this phase, water absorption primarily rises initially, facilitated by the removal of the pericarp (90%), allowing rapid calcium diffusion into the starch within the first 8 hours [31]. In the dry nixtamalization process (ENP), dry milling is employed for size reduction, requiring greater mechanical effort than wet milling used in the TNP. The starch granule undergoes thermal damage during the drying process, and mechanical damage occurs due to dry milling, significantly impacting its functional properties.

Corn kernels are cleaned and selected, and all parts, including the pericarp, are milled together. Achieving an appropriate particle size with a uniform distribution is crucial to prevent flow issues when introduced into the hopper. Dry milling is the commonly used technique for size reduction. A complete diagram of ENP is shown in **Figure 2**.

For the ENP, the selection of the base corn flour is crucial. Characteristics, such as particle size, moisture content, and starch properties, influence the extrusion process and the quality of the final product. Controlling the water content during the nixtamalization process and subsequent grinding is vital. It affects dough consistency, masa hydration, and the overall quality of the final product. Maintaining optimal moisture content in the extrusion process is essential for achieving the desired product attributes. Insufficient moisture can result in a dry texture, while excess moisture can lead to stickiness. So, the ground grain is combined with 0.3% (w/w) lime and has a moisture content ranging between 15 and 30%, depending on the purpose and operating conditions. The conditioned grain is then refrigerated for 12 hours at 5°C [9].

Another point to consider is the use of additives. Some traditional recipes include additives, such as salt, to enhance flavor or other ingredients to modify texture. Balancing these additives is crucial for achieving the desired characteristics in the nixtamalized corn flour. Some products, such as NEX-PLUS a dough conditioner available with or without bleach, facilitate a softer texture during the cooking and steeping of corn [32]. This additive extends the shelf life of the dough and tortillas without compromising taste or aroma. Its key features include preventing brittle edges, reducing sticking, and enhancing elasticity, and it does not interfere with lime.

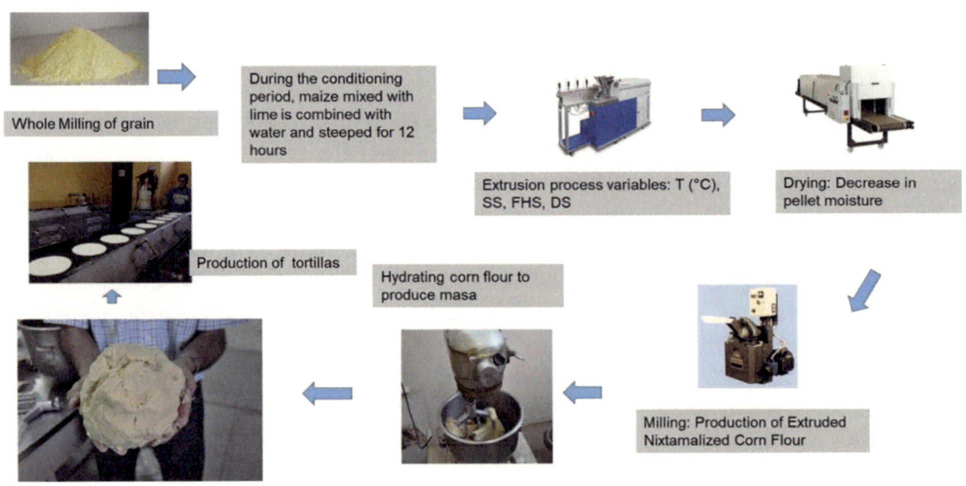

Figure 2.
Extrusion nixtamalization process (the authors). SS: Screw speed. FHS: Feed hopper speed. DS: Extrusion die size.

Discussing the benefits of NEX-PLUS, Enríquez-Castro and Contreras-Jiménez [33] achieved positive empirical results by using 40 to 60 g of NEX-PLUS per 25 kg of nixtamalized corn grain for producing "gorditas" (a type of corn tortillas filled with food or stews) in a commercial food establishment. Although the puffing grade was not measured, the clerk confirmed the achievement of a softer tortilla. The use of this additive is recommended all year long but is particularly beneficial during dry seasons, where almost all of the corn grain is produced through rainfed farming, especially in countries, such as Mexico.

In 2022 and 2023, the northern states, responsible for nearly 90% of corn grain production, experienced limited rainfall, approximately 27 mm according with the National Water Council of México. Consequently, this has led to a reduced corn yield, resulting in the production of older and harder grains, which are then sold to the corn and masa industry [34]. Incorporating specific additives, such as emulsifiers or stabilizers, can influence the rheological properties of the extruded nixtamalized corn flour. These additives may alter viscosity, elasticity, and overall dough behavior. Depending on the desired product, additives, such as emulsifiers, stabilizers, and fortifiers, may be included during the extrusion process. These additives influence the texture, shelf life, and nutritional content of the extruded nixtamalized corn flour.

Proper cooling and drying after extrusion are critical to set the final texture and prevent moisture-related issues. Careful control of these parameters ensures the stability and quality of the extruded product. In both traditional nixtamalized corn flour and extruded nixtamalized corn flour productions, a comprehensive understanding and control of these operating factors and parameters are essential for achieving consistent and high-quality results. The rate at which the extruded dough is cooled post-extrusion is crucial. It affects the setting and final rheological properties of the product. Proper cooling ensures the development of the desired texture [26].

Figure 3 displays two extruded corn pellets obtained with the same formulation but with different water content. This indicates that starch with lower moisture levels may undergo greater dextrinization and a darker coloration, which can affect the quality indicators of the ENCF. This type of flours exhibit reduced particle size,

Figure 3.
A detailed image featuring two varieties of dried extruded corn pellets, each with the same composition but different moisture content (the authors).

facilitating rapid water absorption into molecules. Several researchers attributed increased water absorption capacity in nixtamalized flours to the combined effects of milling and the extrusion cooking process [1, 25]. Thermal processing, a variable contributing to heightened damaged starch content, promotes resistant starch formation due to the loss of crystallinity during extrusion cooking [26]. Temperature fluctuations lead to significant resistant starch formation through hydrolysis and retrogradation. The amylose-amylopectin ratio in corn starch, and the impact of moisture content and temperature, play crucial roles in influencing the functional properties of nixtamalized flour, showcasing a high dependence on these factors in evaluating physical and rheological properties [18].

4.2 Rheological properties

4.2.1 Traditional nixtamalization process (TNP)

Corn flour in TNP is intricately tied to the variables of time and temperature. The duration of the cooking process (8–12 h) and the temperature range (90–96°C) during nixtamalization significantly impact the rheological properties of the resulting dough. These properties, closely linked to viscoelastic parameters, such as elastic modulus and viscous modulus, play a crucial role in determining the quality of the final product [2].

Optimal conditions are paramount to ensuring the proper gelatinization of starch and protein interactions. The corn-to-water ratio during nixtamalization (1:3) directly influences the hydration of starch and proteins, thereby affecting the rheological behavior of the dough. The steeping step is a critical consideration; without proper attention, extended cooking times and prolonged steeping can lead to an increase in gelatinized starch and lower enthalpy values [35]. Allowing the nixtamalized corn masa to rest after preparation is essential for optimal rheological development. This resting period allows for hydration and relaxation of the masa structure. This, in turn, raises the gelatinization temperature and results in a more reorganized starch molecule, contributing to a higher degree of crystallinity retrogradation, shortening, and syneresis [28].

The particle size distribution following the grinding of nixtamalized corn is a key factor influencing dough viscosity and texture. The particle size distribution obtained in the traditional nixtamalization process generates particles ranging from medium to coarse. *This allows for the presence of starch granules in their native structure, in contrast*

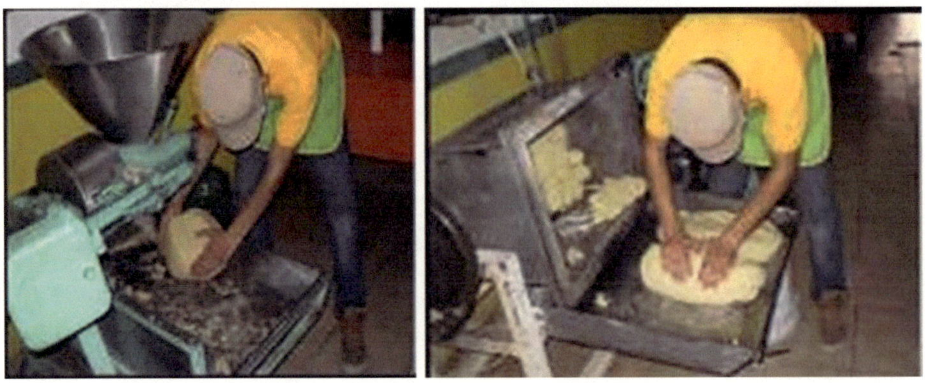

Figure 4.
Detailed image showcasing the milling and conditioning process of corn masa (with Tortillería Aguanaval permitted authorization).

to dry corn milling, where the content of fine particles is higher, potentially causing greater damage to the internal conformation of the starch molecule. A more uniform grind generally leads to improved rheological properties. The theoretical expertise of the operator is particularly crucial in achieving optimal granulometry, as seen in **Figure 4**.

4.2.2 Extrusion nixtamalization process

One of the factors that significantly impacts the process is the rise in temperature, as an elevated temperature entails a greater heat flow through conduction. This increase accelerates the fusion rate, resulting in a consequent reduction in the material's viscosity, and consequently, the starch structure can be modified [9].

Extrusion is performed with low moisture levels (<30%), emphasizing that starch transformation relies more on the mechanical stress inside the extruder than on moisture. Moisture levels below 30% promote granule fragmentation and dextrinization, resulting in heightened extrudate expansion at the cost of crucial functional properties such as rheological, structural, and physicochemical characteristics. Notably, this includes impacts on solubility and water absorption rates [35]. According to Palanisamy et al. [36], an increase in barrel temperature and feed moisture leads to a reduction in the material viscosity within the extruder, directly affecting die pressure. In this scenario, the protein content played a significant role in influencing the response of the extruder.

The screw configuration significantly impacts melting efficiency, with three sections—feeding, transition, and metering. To minimize friction, a simple screw design with a large blade angle is crucial. The length-to-diameter ratio (L/D), measuring the enclosed screw portion, typically falls between 24:1 and 30 or 32:1. However, the appropriate ratio depends on the specific process and application. This ratio correlates with starch granule damage, with a higher L/D ratio causing more thermo-mechanical impact [37]. **Figure 5** presents the sections of a single screw extruder used in the elaboration of foods and cereals.

Three regions are located inside the extruder: transport, increase in volume (swelling), and melting/degradation. This third stage is where the greatest damage to the starch structure is generated. The effect produced by the use of heat combined with the application of mechanical stress weakens the structure of the starch granules, favors gelatinization, and increases the water absorption capacity [39, 40]. Another factor to consider is the moisture content used. Unlike wet milling that uses excess water, dry milling uses little water, which eliminates the protective effect of

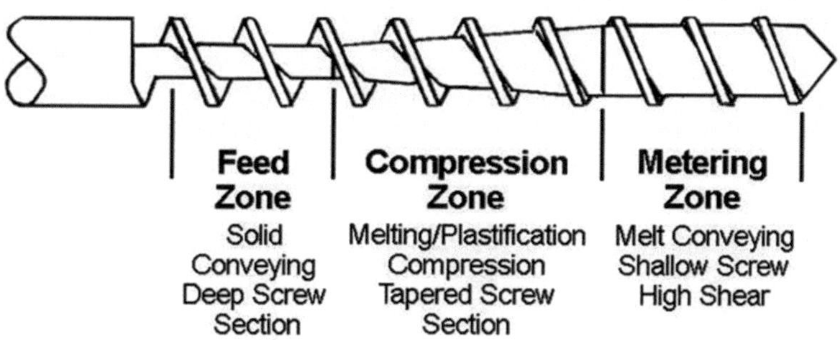

Figure 5.
Sections of a single screw extruder (https://www.globalseafood.org [38]).

water [18]. Understanding and optimizing these operating factors and parameters are essential for achieving the desired rheological properties in both TNP and ENP. This knowledge is fundamental for ensuring consistent quality and meeting specific product requirements.

5. Relationship between nixtamalization and wet extrusion

5.1 Benefits of combining techniques

Utilizing nixtamalization in conjunction with extrusion enhances the bioavailability of calcium and niacin, facilitating the gelatinization of starch and the absorption of these compounds [41]. The extrusion nixtamalization process (ENP) reduces water activity, thereby extending the stability and shelf life of the products. The ENP achieves a more homogeneous and soft texture. Enríquez-Castro et al. [9] conducted a study comparing the texture of tortillas produced through two distinct processes. The measurements were taken at intervals of 2, 24, and 48 h. Surprisingly, there was no discernible difference in texture between the two processes at these time points, indicating that consumers did not perceive variations in palatability between the products. Notably, an important observation emerged after the initial measurement period. Extruded tortillas, when stored in the refrigerator, exhibited greater flexibility compared to their traditionally nixtamalized counterparts. This increased flexibility persisted for up to 7 days, highlighting the potential impact of extrusion on the extended shelf life of tortilla products (unpublished data). Importantly, there was no decomposition of the extruded tortillas within this timeframe, underscoring the significance of extrusion not only in maintaining texture but also in enhancing the overall durability and quality of the product over an extended period.

Extrusion is a rapid and efficient process that enables high-scale production, offering potential benefits to the food industry in terms of cost-effectiveness and enhanced performance. The combination of these two processes allows for the adjustment and personalization of corn masa properties, including texture, flavor, and nutritional profile, in line with consumer preferences. The study conducted by Platt-Lucero et al. [25] on the production of tortillas using ENCF revealed that various factors significantly influence the ultimate quality of the product. Among these factors, particle size, the milling process, and water absorption capacity play crucial roles. It is evident that the starch molecules, when subjected to successive milling, result in increased damage to the inner structure. This damage has a direct impact on the texture of the final product, highlighting the intricate relationship between these processing variables and the overall quality of the tortillas. Another advantage of using this technology is the reduction of mycotoxin levels in corn grain and the potential elimination of other pathogenic agents. Extrusion provides greater control over the production process, resulting in a more uniform corn masa with respect to both physical and nutritional characteristics.

5.2 Difference between HTST process and LTLT process

Extrusion cooking is a high-temperature short-time (HTST) method that utilizes temperature, pressure, and shear force to transform damp starchy and proteinaceous raw materials in low-density products with unique physicochemical characteristic [42].

In extrusion processing, HTST principles can be applied by subjecting the product to a rapid and intense heat treatment for a short duration. The extruder uses high temperatures, often exceeding 100°C, and the product spends only a brief time in the extrusion chamber. This approach is effective for achieving microbial reduction and ensuring food safety while preserving the sensory qualities of the product [43]. Conversely, LTLT principles in extrusion involve using lower temperatures during the extrusion process but extending the processing time. This allows for a more gradual heat application, potentially impacting the flavor, color, and texture of the final extruded product.

Choton et al. [44] conducted a study involving diverse extrusion treatments, revealing a direct correlation between the phenolic compounds content (TPC) and finger millet, as well as sorghum. Their findings indicated that by adjusting processing parameters such as feeder speed, moisture content, screw speed, and temperature, higher TPC levels could be achieved for both flours. Specifically, they reported values of 966.32 mg FE/100 g dw for finger millet and 506.71 mg FE/100 g dw for sorghum. This observation aligns with the conclusions drawn by other researchers, including Neder-Suárez et al. [45], Escalante-Aburto et al. [46], and Menchaca-Armenta et al. [47] who similarly identified the impact of processing parameters on the anthocyanin content in nixtamalized blue corn flour.

5.2.1 Impact on product quality in extrusion

HTST extrusion is commonly favored when maintaining the original characteristics of the product is crucial. It helps retain flavors, colors, and nutritional content, making it suitable for applications where sensory attributes are a priority. The extrusion process does not produce effluent, such as cooking liquor called nejayote. So, valuable compounds are retained, producing a whole corn flour to produce nixtamalized products. Enríquez-Castro et al. [9, 42] proved instrumental information in assessing and contrasting the texture and palatability of tortillas crafted from ENCF in comparison to commercially available corn flour and traditionally processed corn flour. Remarkably, the researchers identified no significant differences between the two products, noting consistent values in terms of texture and rollability across varying time intervals—specifically at 2, 24, and 48 hours.

The LTLT approach in extrusion may be chosen when a longer processing time at lower temperatures is acceptable or desired. It is crucial to recognize that prolonged exposure to heat can exert a significant impact on the quality of the product, potentially leading to variations in flavor profiles or textures. Additionally, when low-shear extrusion dominates the processing conditions within the extruder and is not appropriately established, there is a heightened likelihood of increased starch damage, ultimately yielding dextrinized starch. This underscores the importance of meticulously managing and optimizing extrusion parameters to ensure the desired product attributes and prevent undesirable alterations in starch composition [15].

5.3 Applications in extrusion

HTST extrusion is suitable for a range of heat-sensitive products, such as cereal-based snacks, where preserving the original attributes of the product is essential. LTLT extrusion may find applications in products where a more prolonged heat treatment is acceptable or where achieving specific textural or flavor modifications is desirable. The preconditioning phase of corn, undertaken prior to the extrusion process, proves highly effective in elevating the texture and rheological properties of the flour. This preparatory step is crucial as it sets the foundation for the subsequent transformation into masa, eventually leading to the production of tortillas [2]. The preconditioning stage in corn processing closely resembles the traditional steeping phase in nixtamalization. However, it deviates by utilizing a reduced amount of calcium, effectively eliminating the subsequent risk of contamination or discharge of cooking. Employing this meticulous technique not only optimizes the final texture of the product but also minimizes starch damage, resulting in a superior-quality end product.

The selection between HTST and LTLT principles in extrusion depends on the desired final product goals. HTST is favored for its efficient microbial reduction without compromising product quality, while LTLT is considered when extended processing time at lower temperatures aligns with the desired product characteristics. Adherence to regulatory standards and careful consideration of sensory attributes are paramount in making informed decisions, mirroring best practices in food processing.

In recent years, significant innovations in food extrusion and nixtamalization have emerged, revolutionizing the production landscape. A key focus has been on the development of novel products, cleaner production processes, and energy-efficient operations. One notable innovation lies in the creation of new food products through extrusion and nixtamalization. Researchers and food technologists have explored diverse raw materials and formulations, resulting in a wide array of extruded and nixtamalized products with enhanced nutritional profiles, textures, and flavors. These advancements have expanded the possibilities for offering consumers unique and healthier alternatives.

Cleaner production processes have also become a priority in the field. Efforts have been made to minimize environmental impact by optimizing resource utilization, reducing waste generation, and incorporating sustainable practices. Innovations in cleaning-in-place (CIP) systems, waste recovery, and efficient water and energy usage have contributed to a more environmentally friendly approach in both extrusion and nixtamalization processes.

Energy efficiency has become a central theme in the evolution of these technologies. Researchers have developed new extrusion and nixtamalization systems that maximize energy utilization while minimizing waste. This not only reduces operational costs but also aligns with the global push toward sustainable and eco-friendly practices in the food industry.

One groundbreaking development is the integration of nixtamalization and extrusion as a novel wet process for producing nixtamalized products. This combined approach leverages the benefits of both techniques, enhancing the nutritional quality and sensory attributes of the final products [48]. The synergy between nixtamalization and extrusion allows for more efficient processing, reduced cooking times, and improved textural characteristics, ultimately providing consumers with products that maintain traditional flavors while meeting modern dietary preferences [49].

Extrusion nixtamalization has been managed with a maximum of 30%, resulting in increased starch dextrinization due to the inflicted thermo-mechanical damage. Therefore, tests have been conducted to reduce the speed (rpm) while increasing the moisture percentage. This approach achieves maximum cooking, enhancing protein digestibility. Such extruded products may be ideal for formulating third-generation foods for fish, pets, pasta, and snacks. However, ongoing efforts are directed toward making it more suitable for the production of nixtamalized products. Additionally, implementing this technique in twin-screw extruders allows its use in producing high-protein meat analogs. When handling "chunked" corn kernels, the particle size is larger than that used in traditional nixtamalization. In this manner, broken corn kernels are processed, even though the moisture content can increase up to 45%, and cooking cycles may be repeated up to two or three times. Simultaneously, the equipment gradually reduces the temperature, requiring less water and lime than those used in the traditional process [50].

6. Conclusion

The continuous innovations in food extrusion and nixtamalization reflect a commitment to developing healthier products, adopting cleaner production processes, and achieving energy efficiency. The integration of nixtamalization and extrusion as a wet process further exemplifies the industry's dedication to pushing the boundaries of traditional food processing, offering exciting possibilities for the future of food technology. The integration of extrusion and nixtamalization processes plays a pivotal role in transforming raw ingredients into a diverse range of food products. The choice between dry and wet processes, coupled with meticulous attention to dough composition, influences the quality and characteristics of the final extruded products. Various factors, including temperature, moisture content, and screw speed, significantly impact the efficiency of the extrusion process. Furthermore, in the corn and masa industry, a nuanced comprehension of the HTST and LTLT methods is indispensable. Each method offers specific advantages tailored to the unique requirements and desired outcomes of corn-based products. The intricacies of these techniques directly influence the quality, safety, and shelf life of masa, ensuring that the industry meets stringent standards and consumer expectations. Altogether, these considerations underscore the multifaceted nature of extrusion and nixtamalization in the food industry, highlighting their paramount importance in shaping the sensory, nutritional, and safety attributes of the end products.

Acknowledgements

The authors extend their thanks to all the people of the Food Industry Engineering Department of TECNM Campus Zacatecas Norte who provided support for this work. Also, thanks to Tortillería Aguanaval in Río Grande, Zacatecas, Finally, the authors extend their acknowledgment to TIA Food (Technology in Food Ingredients S.A De C.V) for supporting some of the investigations conducted.

Author details

Carlos Martín Enríquez-Castro[1*], Brenda Contreras-Jiménez[2,3] and Eduardo Morales-Sánchez[2]

1 Tecnológico Nacional de México Campus Zacatecas Norte, Food Industry Engineering Department, México

2 Instituto Politécnico Nacional, Research Center of Applied Science and Advanced Technology, Querétaro, México

3 Engineering Faculty, Universidad Autónoma de Querétaro, University 231, Airport Campus, Querétaro, México

*Address all correspondence to: carlos.ec@zacatecasnte.tecnm.mx

References

[1] Gaytán-Martínez M, Figueroa-CárdenasJD,Morales-SánchezE, Vázquez-LandaverdePA,Martínez-FloresHE. Physicochemical properties of masa and corn tortilla made by ohmic heating. African Journal of Biotechnology. 2011;**10**:16028-16036

[2] Mondragón M, Bello-Pérez LA, Agama-Acevedo E, Betancourt-Ancona D, Peña JL. Effect of cooking time, steeping and lime concentration on starch gelatinization of corn during nixtamalization. Starch-Stärke. 2004;**56**:248-253

[3] Bryant CM, Hamaker BR. Effect of lime on gelatinization of corn flour and starch. Cereal Chemistry. 1997;**74**:171-175

[4] Bello-Pérez LA, Osorio-Díaz P, Agama-Acevedo E, Núñez-Santiago C, Paredes-López O. Chemical, physicochemical and rheological properties of masas and nixtamalized corn flour. Agrociencia. 2002;**36**:319-328

[5] Gutiérrez-Cortez E, Rojas-Molina I, Rojas A, Arjona JL, Cornejo-Villegas MA, Zepeda-Benítez Y, et al. Microstructural changes in the maize kernel pericarp during cooking stage in nixtamalization process. Journal of Cereal Science. 2010;**51**(1):81-88

[6] Briones FC, Iribarren A, Peña JL, Rodriguez RC, Oliva AI. Recent advances on the understanding of the nixtamalization process. Superficies y vacío. 2000;**10**:20-24

[7] Serna-Saldivar SO. Understanding the functionality and manufacturing of nixtamalized maize products. Journal of Cereal Science. 2021;**99**:103205

[8] Ramírez-Jiménez AK, Castro-Muñoz R. Emerging techniques assisting nixtamalization products and by-products processing: An overview. Critical Reviews in Food Science and Nutrition. 2021;**61**(20):3407-3420

[9] Enríquez-Castro CM, Torres-Chávez PI, Ramírez-Wong B, Quintero-Ramos A, Ledesma-Osuna AI, López-Cervantes J, et al. Physicochemical, rheological, and morphological characteristics of products from traditional and extrusion nixtamalization processes and their relation to starch. International Journal of Food Science. 2020;**2020**:5927670

[10] Serna-Saldivar S, Química O. Almacenamiento e Industrialización de Cereales. SA: México DF: AGT, Ed.; 1996

[11] Farías RF, Bustos FM, Moreno YS, Ríos E. Caracterización de harinas comerciales de maíz nixtamalizado. Agrociencia. 2002;**36**(5):557-567

[12] Martínez-Bustos F, Figueroa J.-D.-C., Sanchez-Sinecio F, González Hernández J, De la Luz Martínez J, Ruíz Torres M. Extrusion Apparatus for the Preparation of Instant Fresh Corn Masa. Querétaro, México: U.S. Patent 5558886A, Oct 21, 1995; [Internet] 2023. Available from: https://patents.google.com/patent/US5558886A

[13] Nixtamalization method and device for implementing [Internet]. 2023. Available from: https://patents.google.com/patent/WO2006078153A1/en

[14] Méndez-Albores A, Martínez-Morquecho RA, Moreno-Martínez E, Vázquez-Durán A. Technological properties of maize tortillas produced by microwave nixtamalization with variable alkalinity. African Journal of Biotechnology. 2012;**11**(85):15178-15187

[15] Mercado-Pedraza EB, Morales-Sanchez E, Reyes-Vega M, Gaytán-Martínez M, Ortega Moody JA. Effects of a low-shear transport system on the physicochemical characteristics of nixtamal corn flour. Journal of Food Processing and Preservation. 2014;**38**(3):1330-1338

[16] Guy RCE. Extrusion Cooking Technologies and Applications. Boca Raton, Fla: CRC Press; 2001

[17] Alexander RJ. Corn dry milling: Processes, products, and applications. In: Watson SA, Ramstad PE, editors. Corn: Chemistry and Technology. MN, USA: American Association of Cereal Chemists; 1987. pp. 351-376

[18] Li E, Dhital S, Hasjim J. Effects of grain milling on starch structures and flour/starch properties. Starch-Stärke. 2014;**66**(1-2):15-27

[19] Villada JA, Sánchez-Sinencio F, Zelaya-Ángel O, Gutierrez-Cortez E, Rodríguez-García ME. Study of the morphological, structural, thermal, and pasting corn transformation during the traditional nixtamalization process: From corn to tortilla. Journal of Food Engineering. 2017;**212**:242-251

[20] Gómez MH, Lee JK, McDonough CM, Waniska RD, Rooney LW. Cornstarch changes during tortilla and tortilla chip processing. Cereal Chemistry. 1992;**69**(3):275-279

[21] Adekola KA. Engineering review food extrusion technology and its applications. Journal of Food Science and Engineering. 2016;**6**(3):149-168

[22] Moreno RMM, Figueroa JDC, Santiago-Ramos D, Villa GA, Sandoval SJ, Rayas-Duarte P, Flores HEM. The effect of different nixtamalization processes on some physicochemical properties, nutritional composition and glycemic index. Journal of Cereal Science. 2015;**65**:140-146

[23] Figueroa JDC, Véles-Medina JJ, Hernández-Landaverde MA, Aragón-Cuevas F, Gaytán-Martínez M, Chávez-Martínez E, et al. Effect of annealing from traditional nixtamalization process on the microstructural, thermal, and rheological properties of starch and quality of pozole. Journal of Cereal Science. 2013;**58**(3):457-464

[24] Ruiz-Gutiérrez MG, Sánchez-Madrigal MÁ, Quintero-Ramos A. The extrusion cooking process for the development of functional foods. In: Extrusion of Metals, Polymers and Food Products. London, UK: IntechOpen; 2018

[25] Platt-Lucero LC, Ramírez-Wong B, Torres-Chávez PI, López-Cervantes J, Sánchez-Machado DI, Reyes-Moreno C, et al. Improving textural characteristics of tortillas by adding gums during extrusion to obtain nixtamalized corn flour. Journal of Texture Studies. 2010;**41**(5):736-755

[26] Milán-Carrillo J, Gutiérrez-Dorado R, Perales-Sánchez JXK, Cuevas-Rodríguez EO, Ramírez-Wong B, Reyes-Moreno C. The optimization of the extrusion process when using maize flour with a modified amino acid profile for making tortillas. International Journal of Food Science and Technology. 2006;**41**:727-736

[27] Rojas-Molina I, Gutiérrez-Cortez E, Palacios-Fonseca A, Baños L, Pons-Hernández JL, Guzmán-Maldonado SH, Rodríguez ME. Study of structural and thermal changes in endosperm of quality protein maize during traditional nixtamalization process. Cereal Chemistry. 2077;**84**(4):304-312

[28] Enríquez-Castro CM, Torres-Chávez PI, Ramírez-Wong B, Ledezma-Osuna AI, Quintero-Ramos A, López-Cervantes J, et al. Physicochemical and rheological changes of starch in nixtamalization processes: Extrusion as an alternative to produce corn flour. In: Extrusion of Metals, Polymers and Food Products. London, UK: InTech; 2018

[29] Contreras-Jiménez B, Morales-Sánchez E, Reyes-Vega ML, Gaytán-Martínez M. Functional properties of extruded corn flour obtained at low temperature. CYTA-Journal of Food. 2014;**12**(3):263-270

[30] Martínez-Bustos F, López-Soto M, San Martin-Martinez E, Zazueta-Morales JJ, Velez-Medina JJ. Effects of high energy milling on some functional properties of jicama starch (*Pachyrrhizus erosus L. urban) and cassava starch (Manihot esculenta Crantz)*. Journal of Food Engineering. 2007;**78**(4):1212-1220

[31] Gutiérrez-Uribe JA, Rojas-García C, García-Lara S, Serna-Saldivar SO. Effects of lime-cooking on carotenoids present in masa and tortillas produced from different types of maize. Cereal Chemistry. 2014;**91**(5):508-512

[32] NEX-PLUS: Commercial additives for the dough and corn industry. Available from: https://www.tienda.tiasaalimentos.com.mx/ [Accessed: March 12, 2023]

[33] Enríquez-Castro CM, Contreras-Jiménez BL. (Unpublished data). Study about the way to increase the production and enhance the textural properties of corn masa using Food Process Additive Nex-PLUS (TIASA) in Gorditas Doña Paula, Juriquilla, Querétaro; 2022.

[34] CONAGUA. Gerencia de Planificación Hídrica. Sistema Nacional de Información del Agua (SINA). Available from: https://sinav30.conagua.gob.mx:8080/ [Accessed: March 12, 2023]

[35] Alam SA, Järvinen J, Kirjoranta S, Jouppila K, Poutanen K, Sozer N. Influence of particle size reduction on structural and mechanical properties of extruded rye bran. Food and Bioprocess Technology. 2014;**7**(7):2121-2133

[36] Palanisamy M, Franke K, Berger RG, Heinz V, Töpfl S. High moisture extrusion of lupin protein: Influence of extrusion parameters on extruder responses and product properties. Journal of the Science of Food and Agriculture. 2019;**99**(5):2175-2185

[37] Chung CI. Extrusion of Polymers-Theory and Practice. 2nd ed. Cincinati, Ohio: Hanser Gardner Pubishing, Inc.; 2011

[38] Sections of a single screw extruder [Internet]. 2023. Available from: https://www.globalseafood.org/advocate/extrusion-processing-of-aquatic-feeds-part-1/

[39] von Borries-Medrano E, Jaime-Fonseca MR, Aguilar-Mendez MA. Starch–guar gum extrudates: Microstructure, physicochemical properties and *in-vitro* digestion. Food Chemistry. 2016;**194**:891-899

[40] Alam MS, Kaur J, Khaira H, Gupta K. Extrusion and extruded products: Changes in quality attributes as affected by extrusion process parameters: A review. Critical Reviews in Food Science and Nutrition. 2016;**56**(3):445-473

[41] Ignacio RM, Lannes SCDS. Rheological characterization and texture of commercial mayonnaise using back extrusion. African

Journal of Agricultural Research. 2013;**8**(31):4262-4268

[42] Enríquez-Castro CM, Ramírez-Wong B, Contreras-Jiménez BL, Quintero-Ramos A, de Dios Figueroa-Cárdenas J, Vázquez-Lara F. Effect of extrusion on the crystallinity, viscosity, damage starch, and thermal properties of corn flour, masa, and tortilla. Applied Food Research. 2022;**2**(2):100198

[43] Singh B, Rachna HSZ, Sharma S. Response surface analysis and process optimization of twin sScrew extrusion cooking of potato-based snacks. Journal of Food Processing and Preservation. 2015;**39**(3):270-281

[44] Choton S, Gupta N, Bandral JD, Anjum N, Choudary A. Extrusion technology and its application in food processing: A review. The Pharma Innovation Journal. 2020;**9**(2):162-168

[45] Neder-Suárez D, Lardizabal-Gutiérrez D, Zazueta-Morales JDJ, Meléndez-PizarroCO,Delgado-NieblasCI, Ramírez Wong B, et al. Anthocyanins and functional compounds change in a third-generation snacks prepared using extruded blue maize, black bean, and chard: An optimization. Antioxidants. 2021;**10**(9):1368

[46] Escalante-AburtoA,Ramírez-WongB, Torres-Chávez PI, López-Cervantes J, Figueroa-CárdenasJDD,Barrón-HoyosJM, et al. Obtaining ready-to-eat blue corn expanded snacks with anthocyanins using an extrusion process and response surface methodology. Molecules. 2014;**19**(12):21066-21084

[47] Menchaca-Armenta M, Ramírez-Wong B, Torres-Chávez PI, Quintero-Ramos A, Ledesma-Osuna AI, Frutos MJ, et al. Effect of extrusion conditions on the anthocyanin content, functionality, and pasting properties of obtained nixtamalized blue corn flour (Zea mays L.) and process optimization. Journal of Food Science. 2020;**85**(7):2143-2152

[48] Gulati P, Brahma S, Rose DJ. Impacts of extrusion processing on nutritional components in cereals and legumes: Carbohydrates, proteins, lipids, vitamins, and minerals. In: Extrusion Cooking. Delhi, India: Woodhead Publishing; 2020. pp. 415-443

[49] Kamau EH, Nkhata SG, Ayua EO. Extrusion and nixtamalization conditions influence the magnitude of change in the nutrients and bioactive components of cereals and legumes. Food Science & Nutrition. 2020;**8**(4):1753-1765

[50] Morales-Sánchez E. Nixtamalization in wet milling using extrusion. In: Paper presented at the 9th International Nixtamalization Congress, Juriquilla, Querétaro, México; 28 Sep 2023

Chapter 6

Next-Generation Tools for Nutrition-Inclusive Breeding for Cereals

Sunita Choudhary, Krithika Anbazhagan,
Jana Kholová, Tharanya Murugesan,
Sivasakthi Kaliamoorthy, Keerthi Chadalawada,
Venkata Subramanya Vara Prasad Kodukula, Amol N. Nankar,
Vetriventhan Mani, Maya Chandra, Rashmi Banoriya
and Vincent Vadez

Abstract

Addressing global malnutrition requires improving the nutritional quality of major crops and promoting nutritionally rich crops. However, breeding for improving nutritional traits is challenging, particularly in the absence of rapid and precise phenotyping of these parameters. Quick phenotyping is crucial as it allows breeders to select lines with high nutritional value alongside yield and other important traits while advancing the generations. Traditionally, grain nutritional and quality assessments have relied on wet-lab analytical services, which are slow, costly, and often inaccessible. To overcome these limitations, rapid and cost-effective sensor-based technologies have emerged as a promising solution. Interdisciplinary research combining sensor technology, AI, biochemistry, and crop science has significantly advancing the grain composition analysis, and post-harvest trait evaluation. Tools like near-infrared spectroscopy (NIRS), X-ray fluorescence (XRF), and computer tomography (CT) are increasingly getting utilized to ensure quality standards in trade, nutrition, and food safety. These technologies focus on key traits precisely, time, and cost-effectively, with early findings highlighting their potential for scalable solutions. Such advancements are essential for nutrition-sensitive breeding and improving food safety, quality-based payments for farmers, and supporting global efforts against malnutrition. The swift adoption of these technologies in breeding programs, supported by public-private partnerships, is crucial for sustainable development.

Keywords: malnutrition, rapid grain quality assessment methods, sensor-based technologies, near-infrared spectroscopy, X-ray fluorescence, computed tomography, food safety

1. Introduction

Nutrition-inclusive breeding programs play a crucial role in addressing global malnutrition. In many developing and underdeveloped nations, economic growth has yet to result in better nutritional diets, leaving malnutrition a persistent issue. The effects of climate change further intensify food insecurity in these regions. In India, where over half of the population depends on agriculture for their livelihood and income, the impacts of climate change on food production and nutritional health are especially significant. Without action to promote nutrition-sensitive agriculture, the remainder of the twenty-first century could see a sharp rise in malnutrition. The United Nations designated 2023 as the International Year of Millets (IYM 2023) to raise global awareness of the nutritional and health benefits of millets, while promoting their production as climate-resilient, nutrient-dense cereals. In response, several developing nations have implemented policies and initiatives aimed at addressing the malnutrition crisis by encouraging the cultivation and consumption of these traditionally grown, nutritionally rich crops.

India's "Millet Mission" [1–3] and Kenya's "Blending Policy" [4, 5] are prominent examples of government-led initiatives aimed at addressing nutritional security. In addition, the inclusion of millets as "Nutri-Cereals" has significantly boosted their market value [6]. The Indian Council of Agricultural Research (ICAR), beyond its research and development efforts, has launched several projects in partnership with agrifood business sectors to promote nutri-cereal-based products such as cookies, ready-to-eat foods, and multi-grain flours, helping to reintroduce and popularize millets in the consumer market.

Despite these efforts, more interventions are needed to improve the overall nutritional status of the population. The challenge of ensuring access to adequate nutrition has been a central focus of agricultural research. In response, the Consultative Group on International Agricultural Research (CGIAR; One CG) is aligning crop improvement programs across institutions and regions to drive initiatives that develop nutritionally enhanced crops and advocate for nutrition-sensitive agriculture.

2. Breeding for nutrition in cereals

Crop breeding programs have expanded their focus from merely increasing yield to enhancing the quality of that yield. While boosting the productivity of calorie-dense staple cereals remains a priority, modern breeding efforts also emphasize improving the nutritional content. Achieving significant genetic gains requires an ongoing refinement of methods and strategies. For effective selection and crop improvement, it is crucial to understand both the extent and the dynamics of genotypic and non-genotypic variation, as these factors influence the potential for breeding success and adaptation to changing environmental conditions.

The yield component in cereals has been extensively investigated through research on genotype × environment × crop management (G × E × M) interactions [7–13]. However, a critical aspect that remains underexplored is the nutritional quality of cereals, which similarly responds to variations in environmental conditions and management practices. Nutritional components—especially those pertaining to both grain and stover quality—are crucial for improving human nutrition as well as livestock feed quality, which is mainly crop residue [14–17].

This gap has significant implications for the overall food system, particularly in regions where smallholder mixed crop-livestock systems are prevalent. In sub-humid, semi-arid, and arid tropics and subtropics, where the majority of the world's poorest cereal producers and consumers reside, the nutritional value of cereals plays a dual role: providing sustenance for humans and feed for livestock. The nutritional quality of staple cereals like rice, wheat, and millets directly influence human diets, as these grains often constitute the primary source of calories and essential micronutrients for vulnerable populations. At the same time, the quality of stover and crop residues, which serve as livestock feed, impacts the productivity and health of animals, further contributing to household food security through milk, meat, and other animal products.

In this context, enhancing the nutritional quality of both cereal grains and stover is essential not only for combating malnutrition among human populations but also for sustaining livestock-based livelihoods. As smallholder farmers often rely on both crops and livestock, improvements in crop nutrition can lead to a virtuous cycle of better livestock health and productivity, which in turn supports improved human nutrition through diversified diets and economic stability.

Addressing the nutritional quality of cereals thus requires a broader, integrative approach, combining advanced breeding techniques, agronomic practices, and policy interventions aimed at promoting nutrition-sensitive agriculture. This involves understanding the differential response of nutritional traits across environments and management practices and ensuring that the benefits of crop improvement reach smallholder farmers and their communities, who are most vulnerable to the impacts of malnutrition and food insecurity.

Evaluating cereals for nutritional quality traits within breeding programs poses considerable challenges due to the reliance on specialized analytical laboratories, which require advanced equipment, specific reagents, and skilled personnel [18–20]. The high costs associated with these resources, coupled with limited accessibility, further constrain breeders from routinely incorporating nutritional assessments. Traditional wet-chemistry techniques, while precise, are both time-intensive and laborious, adding to the complexity of the selection process.

This creates a significant bottleneck, as breeders often face the need for rapid decision-making to advance crop generations within a short timeframe. Considering that the development of a new crop variety can span 5–10 years, depending on the species and available resources, delays in nutritional phenotyping can prolong the process and impede the timely release of improved varieties [21–24].

Breeders also face several other challenges when working to improve the nutritional quality of crops. These include the complex genetic regulation of nutritional traits, limited genetic diversity within crop germplasm, negative linkages between nutritional and agronomic traits, environmental influences on trait expression, and varying consumer preferences and acceptance. These factors collectively hinder the development of crop varieties with enhanced nutritional profiles. However, many of these obstacles can be mitigated through the use phenotyping technologies that bridge multiple disciplines, enabling large-scale and multiple evaluation and testing of traits [25–27] when nutrition phenotyping is a significant bottleneck in terms of cost, time and precision.

To accelerate crop selection for nutritional improvement and product development, the integration of advanced phenotyping technologies is essential. These technologies, when coupled with rigorous metrics to quantify precise nutritional parameters, offer the potential for high-throughput evaluation and faster breeding cycles. This approach will be critical for achieving both the quantity and speed required to meet global demands for nutritionally superior, climate-resilient crops and accelerate genetic gains [28–32].

3. Market potential for nutritional products and quality-based payment

Introducing new crop varieties with enhanced nutritional traits presents challenges, particularly concerning consumer acceptance, market demand, and cultural preferences. Gaining consumer trust and fostering education about the benefits of these new products is critical for their successful adoption. In developing countries, however, the adoption of such improved cultivars is often slow and difficult, especially when enhanced nutritional traits do not result in direct economic benefits for farmers. The absence of economic incentives tied to the trade of bio-fortified crops hampers their widespread adoption, as smallholder farmers, who form the backbone of agricultural systems in these regions, are often excluded from premium markets. This is due to the unregulated, localized nature of grain trade, which depends heavily on intermediaries and lacks standardization. Consequently, farmers are disincentivized from investing in seed of nutritionally improved varieties.

Additionally, the prevalent issues of food adulteration and contamination further exacerbate food safety concerns, limiting access to high-quality dietary sources. In these developing regions, assessing grain quality and detecting contaminants traditionally relies on costly and time-consuming wet-lab analyses, which remain inaccessible to many underprivileged communities. The development of sensor-based technologies has the potential to revolutionize this landscape by providing quick, reliable, and cost-effective measurements of grain quality. These technologies, adaptable for use both in the field and in markets, offer rapid, environmentally friendly solutions that are tailored to decentralized market systems (**Figure 1**). By empowering smallholder farmers, vendors, and crop improvement programs alike, these innovations can facilitate better access to premium markets and enhance food safety, ultimately supporting improved livelihoods and nutrition across developing countries.

4. Need of new generation technologies for nutrition

Currently, there are only few techniques available for quick, reliable, and cost-effective field measurements of agricultural produce qualities that fit the context of trading systems in developing countries and allow smallholder farming

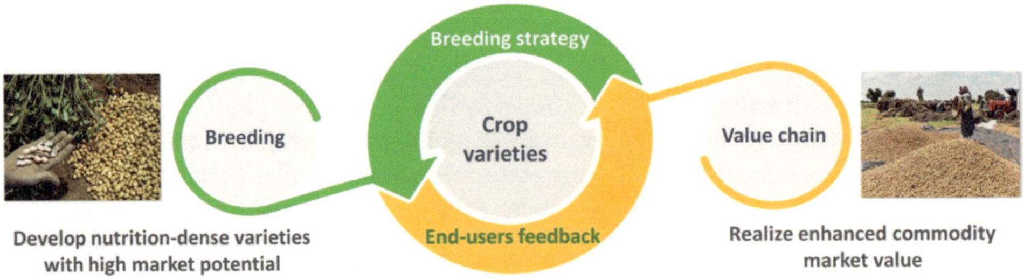

Figure 1.
Pictorial representation of the interrelationship between agrifood research and industrial sectors. The crop breeders and value chain actors are the key drivers for nutrition security-related programs. They also operate in the potential spaces for technological interventions such as integrating sensor-based crop quality assessment in their processes.

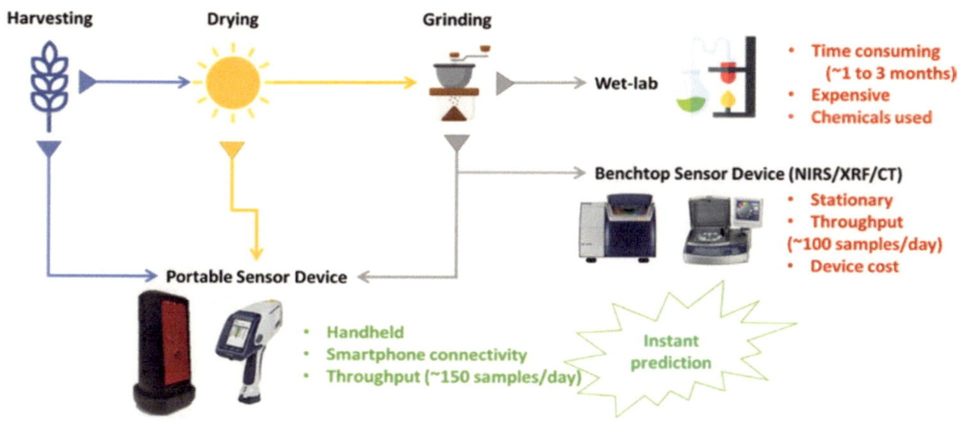

Figure 2.
A graphical representation of the potential of sensor-based technologies to be integrated in agri-research and industrial sectors.

communities to reap the benefits linked to the quality of their produce [33]. A similar situation is observed in the crop improvement programs; in this case, the laboratory facilities may be available, but these are usually not rapid enough to facilitate efficient selection within the breeding populations. For this purpose, research on translating sensors for rapid, robust, cost-effective, yet environmentally friendly grain quality analyses has been initiated (**Figure 2**) [25, 34]. Also, mobilizing sensor-based solutions like handheld devices is critical for catering to the needs of the grain-related value chain players and crop improvement programs that are primarily guided by market-demanded traits, and probably, it can trigger quality-based payment in future.

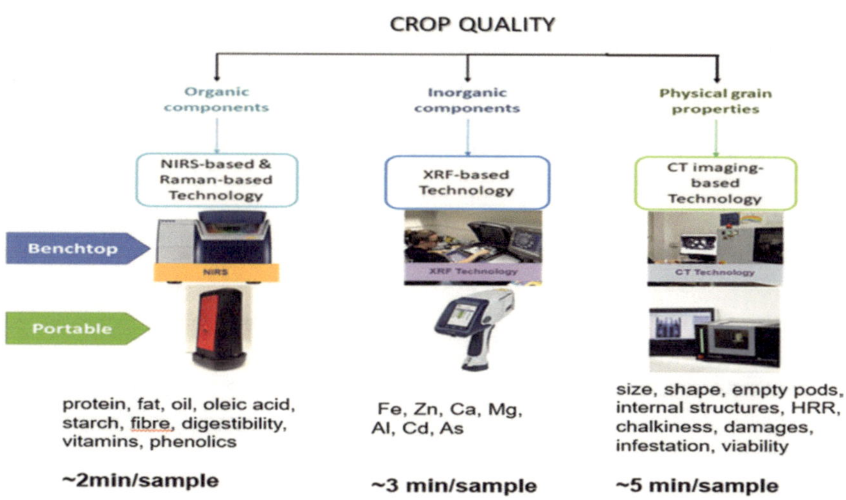

Figure 3.
A brief overview of benchtop and portable sensors (near-infrared spectroscopy, NIR; X-ray fluorescence, XRF; and X-ray computed tomography, CT) that can be deployed in breeding programs as well as value chain pipelines for rapid crop quality assessment.

NIR Instrument type	Benchtop	Portable
Application:	Screening and identification high and low nutritional composition for breeding selection and research	Rapid assessment of high nutritional samples in field and in crop value chain
Spectral range	Wide, for example, NIRS 400–2500 nm	Narrow, for example, NIRS 1100–2500
Analyses time:	less than 1 minute	1 minute
Throughput	250 whole grain samples per day 100 flour samples per day	200 whole grain samples per day 150 flour samples per day
Portability	Low and tedious	High and easy
Connectivity	Desktop computer	Mobile phone or laptop
Prediction Precision	High*	Low* (compared to benchtop but significantly accurate)
References	[32, 35, 36]	[34, 35, 37, 38]

**Depending on trait, calibration method, crop, and trait variation.*

Table 1.
An example of a benchtop and a portable near-infrared (NIR) spectrometer, X-ray fluorescence (XRF), and computed tomography (CT).

4.1 Benchtop vs. handheld sensors

The mobility of sensor technology solutions is a critical attribute, particularly when considering the needs of grain-related value chain players in the decentralized market context of developing countries. Benchtop sensors, while accurate and reliable, lack the flexibility and portability required to meet the demands of field operations and decentralized markets. In contrast, mobile sensors offer the advantage of portability and on-the-go measurements, making them well-suited for use in remote locations and field settings. This mobility is essential for catering to the needs of crop improvement programs, which are increasingly guided by market-demanded traits (**Figure 3**, **Table 1**). By providing real-time data and analysis capabilities directly in the field, mobile sensors empower stakeholders to make informed decisions swiftly and efficiently, ultimately enhancing the productivity and resilience of agricultural systems in developing countries.

5. Types of sensors

Several types of sensors are used in various fields like material science, pharma, medical, agrifood and feed industry, and agriculture. In this paper, sensors of near-infrared spectroscopy (NIRS), X-ray fluorescence (XRF), computed tomography (CT), and Raman spectroscopy are discussed (**Figure 3**) with a special focus on NIRS.

5.1 Near-infrared spectroscopy (NIRS)

NIRS operates by exposing samples to near-infrared light, which interacts with the molecular vibrations of the sample, producing a unique spectral fingerprint. The NIRS machine detects the intensity of light at various wavelengths after it has passed through or been reflected by the sample [39]. By comparing these absorption patterns to a pre-established calibration model, which correlates known chemical compositions of samples with their corresponding near-infrared absorption spectra, the NIRS machine can estimate the organic composition of nutritional traits. This technology has been successfully applied to a range of crops for organic traits like proteins, starch, fiber, fatty acids, amino acids, and fats. For instance, it has been used to estimate protein and amylose content in rice [40], lignan glucoside content in sesame [41], and oil content in safflower [42]. Both benchtop and handheld NIRS equipment are available, with benchtop models typically covering a wider wavelength range from visible light (400–2500 nm), while handheld models capture a narrower range (1100–2500 nm). The wavelengths of the sensors vary across different manufacturers of portable sensors [43].

5.2 X-ray fluorescence (XRF)

XRF, on the other hand, relies on the principle of X-ray excitation of atoms in the sample, causing them to emit characteristic fluorescent X-rays. These X-rays are then detected and analyzed to determine the elemental components like metals. XRF works with X-ray energies ranging from a few hundred electron volts to several kilo-electron volts. X-ray ray typically utilizes the band below 400 nm and are used for the prediction of inorganic nutritional traits like iron (Fe), zinc (Zn), calcium (Ca), and contaminants like aluminum (Al), arsenic (Ar), and lead (Pb). Benchtop standard XRF equipment is used to predict the inorganic composition of cereal grains, such as wheat and rice [44, 45]. The standard models were used to calibrate the position of the sample, particularly inorganic components like metals [46, 47]. The effectiveness of calibrations from both systems was evaluated by comparing the coefficient of determination [2], regression slope, and root mean square error (RMSE).

5.3 Computed tomography (CT)

CT imaging involves taking a series of X-ray images of a sample from different angles and using computer algorithms to reconstruct a 3D image. By analyzing the attenuation of X-rays as they pass through the sample, CT can provide detailed information about the internal structure, density, and morphology of the sample, making it particularly useful for non-destructive analysis of materials and biological specimens. Each of these techniques operates within specific spectral ranges, and CT imaging utilizes X-rays with energies in the range of tens to hundreds of kiloelectron volts [48]. The CT applications prove invaluable for quickly evaluating basic grain properties—such as size, shape, and damage—especially in crop species with grains covered by a hard-to-remove shell, like groundnut and rice [49, 50]. The CT equipment is also available in portable size.

5.4 Raman spectroscopy

Raman spectroscopy is a technique used in analytical chemistry and materials science to study vibrational, rotational, and other low-frequency modes in a system. It involves shining a laser light onto a sample and measuring the scattered light. The shifts in energy, known as Raman shifts, provide information about the chemical composition, crystal structure, and molecular interactions within the sample. The RS is a spectroscopic method that indirectly measures the vibrational states of samples based on the polarity of chemical bonds [51, 52]. Due to its insensitivity to water and fewer overlapped bands, RS is suitable sensor for accurate qualitative and quantitative assessment of liquid samples where NIR measurements are restricted [52]. Also, in high moisture-containing samples of grains and oilseeds, the detection of fungal and mycotoxin is easier with RS by minimizing inferences from residual components. For instance, Yuan et al. [53] utilized surface-enhanced Raman spectroscopy (SERS) to quickly detect the mycotoxin deoxynivalenol in corn, kidney beans, and oats. The SERS was also used to identify aflatoxin in maize. [54] have used RS to detect carbohydrate, fiber, carotenoid, and protein content in maize kernels.

6. Recent developments with sensors

6.1 Near-infrared spectroscopy (NIRS)

NIRS has the potential to be a highly effective tool for measuring traits associated with organic grain composition, provided the calibrations are meticulously developed and rigorously validated. For industrial applications, single-species NIR calibrations are generally preferred over multispecies calibrations because they offer greater accuracy and precision [33, 41, 42]. However, producing single-species calibrations can be challenging in certain cases, such as with "minor millet" (including pearl, finger, Kodo, Proso, Foxtail, Little millet, Sorghum, and Teff) or "orphan" legumes (such as horsegram, lablab, cowpea, mothbean, and adzuki bean), where limited diversity may hinder the development of robust calibrations. In these situations, multiple-species calibrations are used instead (**Figure 4**). For example, the performance metrics of a single-cereal equation for sorghum, built using 96 samples, were compared to those of a multiple-cereal equation constructed with 62 samples of

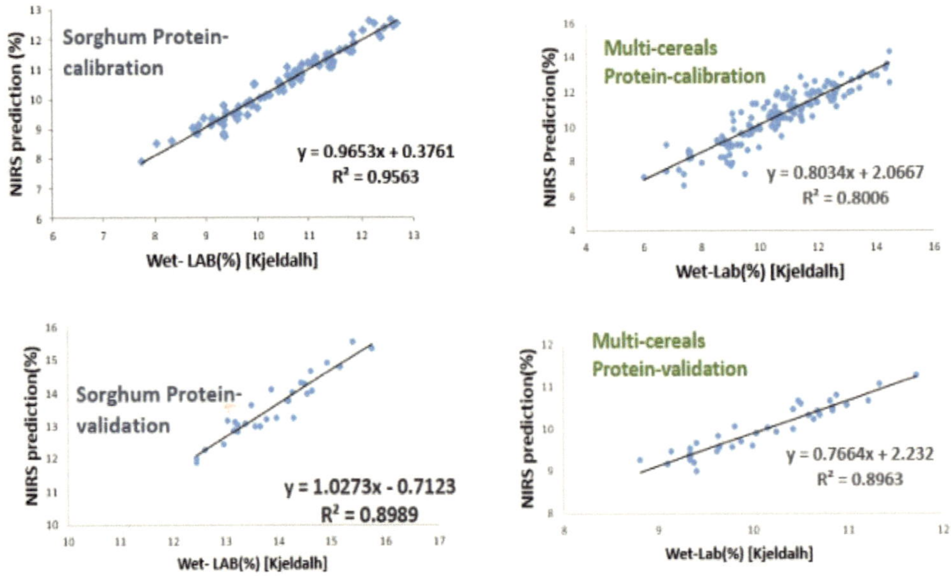

Figure 4.
Matrix of XY-scatterplots depicting the NIR-based grain protein content prediction models along with metrics for single species (sorghum) and multiple species (sorghum, pearl millet, finger millet, foxtail millet). The top panels represent the calibration or training dataset, and the lower panel represents the cross-validation dataset.

sorghum, 26 samples of pearl millet, 20 samples of finger millet, and 20 samples of foxtail millet, using various calibration methods (see **Figure 5**). The sorghum-specific calibration, which used a sufficient sample size (96) and a broad range of variation (protein: 6.7–14.2%; fat: 1.5–5.4%), predicted sorghum protein and fat more accurately (see **Figure 4**). However, the multiple-species prediction can still be applicable for sorghum, with calibration results showing R^2 values of 0.8 for protein and 0.92 for fat, and RMSE values of 0.78 for protein and 0.66 for fat. Validation metrics include R^2 values of 0.85 for protein and 0.78 for fat, RMSEP values of 0.4 for protein and 0.22 for fat, and slopes of 1.3 for protein and 0.88 for fat. This approach might be preferable for crops where the sample size and range of variation are insufficient for a single-species calibration.

6.2 X-ray fluorescence (XRF)

XRF-based technologies have traditionally been employed in heavy industrial applications, with their use in crop research only becoming more prominent in the past decade. Research on pearl millet has demonstrated significant genetic variability for iron and zinc (30–140 mg kg^{-1} Fe and 20–90 mg kg^{-1} Zn), which can be leveraged to develop high-yielding cultivars rich in these nutrients. However, previous breeding programs primarily focused on yield improvement, resulting in released cultivars and hybrids with lower Fe and Zn content [55]. Recognizing this opportunity, the All India Coordinated Research Project on Pearl Millet (AICRP-PM) of the Indian Council of Agricultural Research (ICAR) established a threshold of minimum iron (42 mg kg^{-1}) and zinc (31 mg kg^{-1}) levels for release of national varieties of pearl millet. This policy promoted biofortified pearl millet in India but also necessitates rapid assessment methods for selecting high Fe and Zn lines. Since then, XRF technology has been extensively used to screen materials, contributing significantly to alleviating malnutrition [56]. Similarly, finger millet, which is rich in calcium, required breeding efforts to select high Ca content lines. The calibration for estimating Ca content in finger millet grains using benchtop standards is currently being optimized at ICRISAT. Portable XRF is a non-destructive method that allows comparison with high-precision techniques like ICP-MS on the same sample, as demonstrated in rice

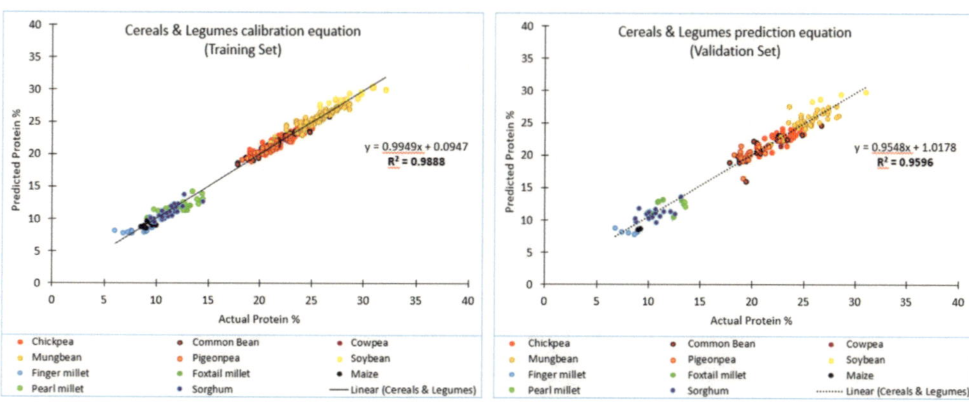

Figure 5.
Example of NIR-based protein prediction model for grain flour samples of multiple crops (cereals and legumes). The XY scatterplot of left presents the metrics of calibration or training dataset and the scatterplot of right presents the metrics for the cross-validation used for determining the robustness of the prediction model.

[45]. This report presents portable XRF measurements for As, Mn, Fe, Ni, Cu, and Zn in rice samples, comparing them with ICP-MS results for the first time. While these sensors need systematic testing on a broader range of materials, the example of pearl millet clearly illustrates that XRF application is feasible, provided the technology is rigorously tested and further developed for estimating the elemental composition of crop grain material.

6.3 Computed tomography (CT)

It has been widely used in the medical industry as well as in heavy-industry applications [57, 58]. Again, the applications for phenotyping of plant-related material are a relatively recent development [59–61]. Despite this, CT-enabled information could quickly become a game-changing factor that can enable critical information for crop improvement programs as well as for estimation of crop value in the market systems (rapid in-shell estimates such as grain weight, size and shape, damages, and shelling ratio, see **Figure 6** for example). CT has been tested to address such demand from rice and groundnut crop improvement programs where the time needed for grain threshing becomes a significant barrier in the grain evaluation during the selection process [50]. Similarly, the groundnut processing industries (e.g., Greenforest Foods Ltd., Kenya) indicated that rapid access to this information could overcome the market barrier between the small-scale producers and the large groundnut processing industries. This technology is currently being systematically tested to address these particular needs.

6.4 Raman spectroscopy

As an analytical tool, RS can provide molecular insight into mycotoxin [62, 63]. Several studies showed promising results for rapid screening of mycotoxin-contaminated grains and oilseeds and their products by Raman. Raman spectroscopy is sensitive enough to detect chemical functional groups of mycotoxin compounds and derivatives. Therefore, we expect to be able to characterize mycotoxin molecules through the molecular fingerprint of Raman and further correlate Raman information with the levels of mycotoxin contamination, as demonstrated in the quality control of cereal products [62]. Raman spectroscopy can be used to analyze the

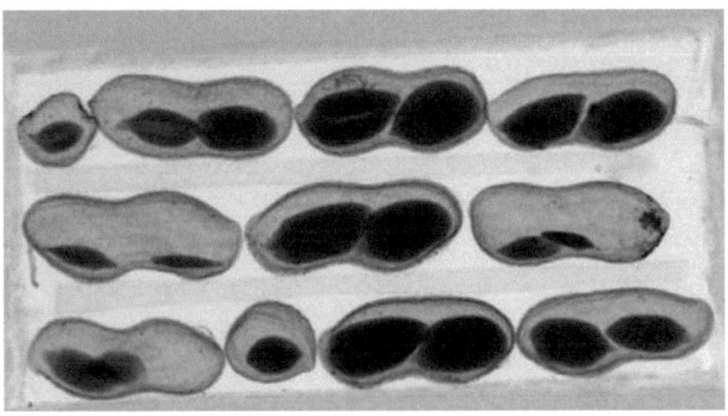

Figure 6.
A two-dimensional X-ray computed tomography image of peanuts showing variations in grain filling that translates into key traits for selection in breeding pipeline as well as for deciding the price at agri-commodity procurement centers i.e., shelling percentage and kernel weight.

chemical composition of agricultural products such as grains, fruits, and vegetables. It is increasingly recognized for its effectiveness in assessing plant health, crop quality, and species, as well as detecting plant maturity, ripeness, freshness, nutritional content, presence of contaminants or pesticides, abiotic stresses, and diseases [64–66]. It supports rapid phenotyping and digital plant selection, facilitating early detection of bacterial infections, fungal diseases, insect invasions, and other pathogens in greenhouses and fields [64, 66, 67].

7. NIRS as emerging powerful technology for nutritional breeding pipeline

Technological tools are available to greatly enhance the nutritional value of breeding programs. However, enhanced nutrition in any breeding program might only be achieved if the screening of nutritional traits is handy, economical, and fast. NIRS is one of the most proven tools to measure organic grain and stover composition like protein, fat, starch, amino acid, polyphenols, phytic acid, amino acid, *in vitro* digestibility, metabolizable energy, fiber content, etc., in large sample size where chemical analysis is time consuming and expensive [33, 42, 68]. NIRS has been advanced to sensitize the variations across year, location, variety, time of sowing, etc. [69]. NIRS technology holds promising power for nutrition research by accelerating the data collection in crop improvement and understanding the physiology of nutritional traits at low production cost to effectively select the best for human and animal nutrition.

8. Physiological basis of nutritional trait: Case study in sorghum

Sorghum, a versatile crop cultivated worldwide for food, feed, and biomass, holds significant importance for ensuring food security, particularly in vulnerable semi-arid agricultural regions. Despite its crucial role, much of the genetic diversity of sorghum, preserved in the ICRISAT gene bank in India with approximately 40,000 accessions, remains underexplored, particularly regarding variations in grain and stover nutritional composition. The grain nutritional composition of 3000 sorghum accessions with NIRS-based benchtop reveals intriguing insights in **Figure 7**. The density gradient map of each group of accessions, including historic, exotic collection, and ICRISAT Sorghum collections, exhibited distinct positions along this relationship. Noteworthy is the trend observed in historical collections, which typically displayed higher fat content and lower protein content than other groups. Also, recently developed materials in breeding programs are able to increase the protein content and reduce fat. These findings shed light on the diverse genetic variation for nutritional traits available for sorghum germplasm and underscore the importance of further exploration and utilization of this genetic diversity in breeding programs aimed at enhancing nutritional traits.

Accessing and leveraging the vast variability in crop traits is imperative for driving tangible improvements through crop breeding endeavors. Beyond merely identifying this variability, a deeper understanding of the underlying biological processes and their interactions with the environment is essential. In one of the studies ([1]; unpublished), the stover and grain quality of 18 sorghum genotypes across various factorial agronomic treatments (water stress, planting density, and nitrogen fertilization). NIR-based leaf, stem, and panicle quality assessments during early growth stages showed prediction potential for grain nitrogen content as well as stover quality at crop

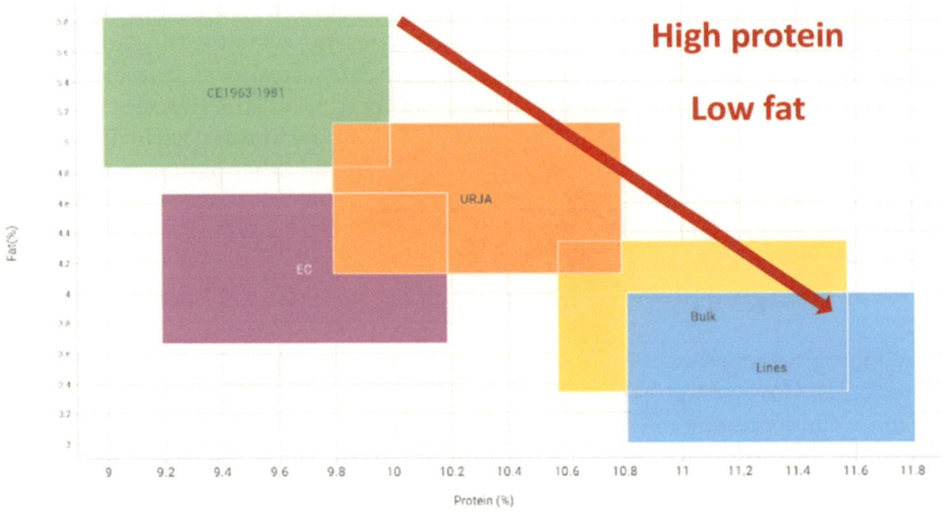

Figure 7.
Graphical representation of the negative relationship between grain protein and fat content in sorghum germplasm collections of ICRISAT.

maturity (**Figure 8**). In addition, significant correlations between leaf senescence and stover quality suggest the potential for indirect field assessments of stover quality based on visual scoring. The interconnectedness of various plant traits and their implications for crop nutritional quality suggests avenues for accelerating breeding efforts through indirect methods.

Furthermore, functional links between plant allometry at maturity and grain nutrient composition offers insights into predicting grain nutritional content based

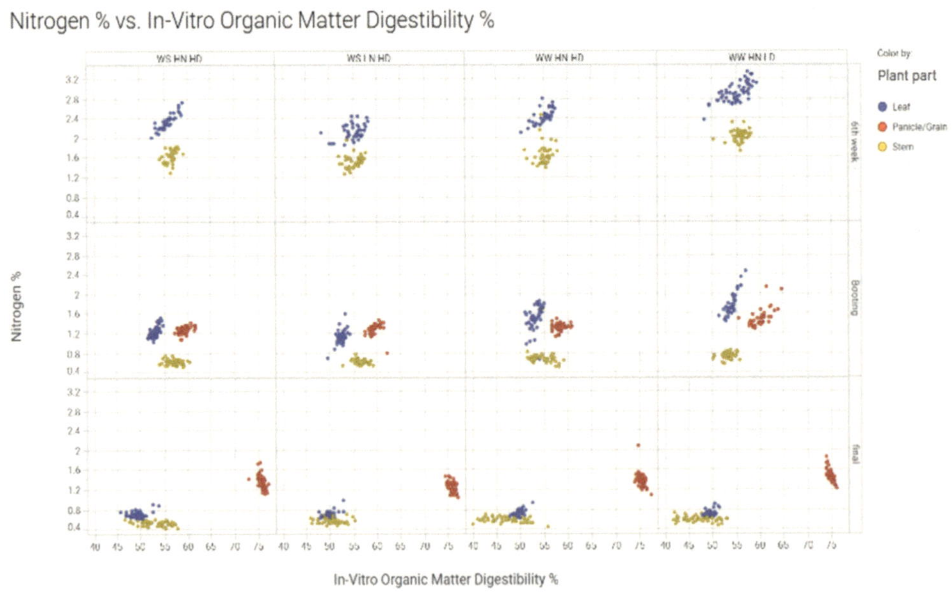

Figure 8.
Matrix of XY-scatter plots indicating relationships between grain nitrogen content and in vitro organic matter digestibility in different plant parts depicted through different colors (leaf: blue, panicle and grain: red; stem: yellow).

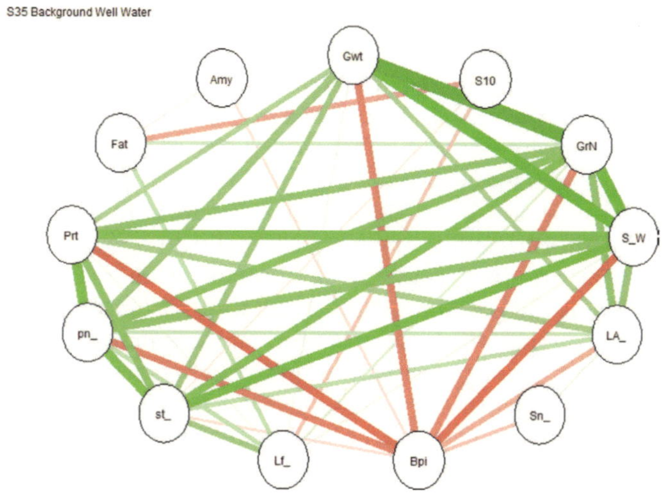

Figure 9.
Correlation networks depicting the interrelationships between different agronomic and nutritional traits (Gwt = grain weight [g plant-1], GrN = grain number per panicle, S_W = stover weight [g plant-1], LA_ = canopy leaf area [cm2], Sn_ = senescence score [0–100%], Bpi = Biomass partitioning index, Lf_ = Leaf in-vitro organic matter digestibility [w/w%], st_ = Stem in-vitro organic matter digestibility [w/w%], pn_ = panicle or grain in-vitro organic matter digestibility [w/w%], Prt = grain protein content [w/w%], Fat = grain fat content [w/w%], and Amy = grain amylose content [w/w]. Green lines and red lines in correlation networks represent positive and negative correlations respectively. Strength of each line indicates the strength of correlation between the pairs. Values within each line in a correlation network represent correlation coefficients.

on plant parameters ([32], unpublished, **Figure 9**). Additionally, the observed synergistic associations between stover productivity and quality, as well as grain quality, as illustrated in **Figure 9**. The correlation network analysis revealed interrelationships among various parameters, including grain weight (both per plant and per hectare), grain size, grain number, stover weight (per plant and per hectare), maximum leaf area (at booting stage), leaf area index, senescence score, biomass partitioning index (BPi), and the nutritional composition of leaves, stems, and panicles. This method aids in visualizing and identifying patterns, revealing key variables and potential causal relationships within complex datasets. For instance, biomass partitioning (Bpi) is negatively associated with grain weight (Gwt), grain numbers (GrN), stover weights, protein content (Prt), and panicle weight. Additionally, grain weight (Gwt) is positively associated with grain numbers (GrN) and stover weight, while negatively associated with biomass partitioning (Bpi). Furthermore, protein content (Prt) is positively associated with stover quality, grain numbers, panicle weight, and grain weight, and negatively associated with biomass partitioning (Bpi). Moreover, stover quality is positively associated with protein content, grain weight, grain numbers, and stem weight, and negatively associated with biomass partitioning (Bpi). These findings underscore the importance of integrating multidimensional data from field experiments with advanced analytical techniques to inform and expedite crop improvement strategies aimed at enhancing nutritional qualities.

9. Conclusion

Nutrition-inclusive breeding and value chains necessitate adoption of reliable, robust, and applicable tools for rapid phenotyping of crop nutritional composition.

Especially for making informed decisions on the choice of crop, farm management, product development, and processing, delivery for human as well as animal nutrition and well-being. Advances in instrumentation, computational power, digital networking, and internet access can be tailored to the agricultural sector and further optimized for different contexts and stakeholders. This book chapter focuses on the sensor-based technologies for crop quality assessment in crop breeding programs, but this can further be connected upstream and downstream with remote sensing and agrometeorological information, econometric modeling, socio-cultural preferences, commodity pricing and trading, marketing and certification, integrated farm advisory tools, etc. Agricultural development requires multidimensional systemic interventions considering diverse players, constraints, and opportunities. Sensors are one such tool that can be explored and integrated at various stages of farming to enable stakeholders to measure quantity as well as quality, which is essential for streamlining nutrition-related breeding and crop value chains. Such efforts are imperative to accelerate the development of nutritionally dense grain-crop products, particularly for the most vulnerable agricultural systems and stakeholders. Incorporating sensor-based technologies like NIRS and XRF into regular breeding programs and crop value chains holds promise for accelerating nutrition selection alongside productivity. At the same time, it is essential to understand the complexities of nutrition physiology and cascade mechanisms for fully exploiting these tools in breeding endeavors. Establishing global networks that unite experts from various fields is crucial for developing and implementing sensor-based technologies in agrifood research and industry. This also involves co-development of sensors (i.e., testing, calibration, and validation of sensors) with different stakeholders and for the development of nutritionally dense safe products tackling malnutrition.

Author details

Sunita Choudhary[1*], Krithika Anbazhagan[2], Jana Kholová[1], Tharanya Murugesan[1], Sivasakthi Kaliamoorthy[1], Keerthi Chadalawada[3], Venkata Subramanya Vara Prasad Kodukula[2], Amol N. Nankar[4], Vetriventhan Mani[1], Maya Chandra[5], Rashmi Banoriya[5] and Vincent Vadez[6,7]

1 International Crops Research Institute for Semi-Arid Tropics (ICRISAT), Patancheru, Telangana, India

2 International Livestock Research Institute, Patancheru, Telangana, India

3 Bharathidasan University, Tiruchirappalli, Tamil Nadu, India

4 Institute of Plant Breeding, Genetics and Genomics (IPBGG), University of Georgia, Tifton, Georgia, USA

5 Dr. Rajendra Prasad Central Agricultural University, Samastipur, Bihar, India

6 IRD (Institut de Recherche pour le Développement), Montpellier, France

7 CERAAS (Centre d'Etude Régional pour l'Amélioration de l'Adaptation à la Sécheresse), Thiès, Senegal

*Address all correspondence to: sunita.choudhary@icrisat.org

References

[1] Choudhury PP. Odisha millet mission, another nutritious cereal for healthy living. Agriculture and Food: E-Newsletter. 2023;**5**(2):223-225

[2] Kumari P, Thakur A, Sankhyan NK, Singh U. Millet production and consumption in India and their nutritional aspects. Just Agriculture. 2023;**3**(5):46

[3] Verma HP, Verma A. Chhattisgarh millet mission. Just Agriculture E-Magazine. 2023;**3**(5):67-68. Available from: https://justagriculture.in/files/newsletter/2023/january/91.%20Chhattisgarh%20Millet%20Mission.pdf

[4] Melesse MB, Tessema YM, Manyasa E, Hall A. An enabling environment for the national flour blending policy: A food systems analysis. In: Breisinger MK, Mbuthia J, Njuki J, editors. Part 6: Toward more Sustainable Food Systems, Chapter 16 Food Systems Transformation in Kenya: Lessons from the Past and Policy Options for the Future. Washington, DC: Intl Food Policy Res Inst; 2023. pp. 409-432. DOI: 10.2499/9780896294561_16

[5] Poole N, Donovan J, Kariuki S, Rutsaert P, Ibba MI, Bentley A. Flour blending can mitigate food insecurity and economic stress. Global Food Security. 2024;**41**:100758

[6] WPI. Wholesale Price Index. 2018. Available from: www.eaindustry.nic.in

[7] Seyoum A. Multi-environment evaluation and genotype x environment interaction analysis of sorghum [*sorghum bicolor* (L.) Moench] genotypes in highland areas of Ethiopia. Journal of Environmental Geology. 2021;**5**(3):1

[8] Motlhaodi T, Bryngelsson T, Chite S, Fatih M, Ortiz R, Geleta M. Nutritional variation in sorghum [*Sorghum bicolor* (L.) Moench] accessions from southern Africa revealed by protein and mineral composition. Journal of Cereal Science. 2018;**1**(83):123-129

[9] Wirnas D, Sopandie D, Tesso T. Genotypes X environment interaction effect on nutritional quality of sorghum lines in Indonesia. Ekin Journal of Crop Breeding and Genetics. 2015;**1**(2):26-31

[10] Cakmak I. Enrichment of cereal grains with zinc: Agronomic or genetic biofortification. Plant and Soil. 2008;**302**:1-7

[11] White PJ, Broadley MR. Biofortification of crops with seven mineral elements often lacking in human diets–iron, zinc, copper, calcium, magnesium, selenium and iodine. New Phytologist. 2009;**182**(1):49-84

[12] Graham RD. Breeding for nutritional characteristics in cereals. Advances in Plant Nutrition. 1984;**1**:57-102

[13] Swamy BM, Marathi B, Ribeiro-Barros AI, Ricachenevsky FK. Development of healthy and nutritious cereals: Recent insights on molecular advances in breeding. Frontiers in Genetics. 2021;**12**:635006

[14] Blümmel M, Duncan AJ, Lenné JM. Recent advances in dual purpose rice and wheat research: A synthesis. Field Crops Research. 2020;**15**(253):107823

[15] Upadhyaya HD, Vetriventhan M, Are AK, Azevedo VCR, Wang YH. Sorghum Germplasm for enhanced productivity and nutrition. In: Tonapi VA, Talwar HS, Are AK, Bhat BV,

Reddy CR, Dalton TJ, editors. Sorghum in the 21st Century: Food – Fodder – Feed – Fuel for a Rapidly Changing World. Singapore: Springer; 2020. pp. 101-123. DOI: 10.1007/978-981-15-8249-3_5

[16] Parissi Z, Irakli M, Tigka E, Papastylianou P, Dordas C, Tani E, et al. Analysis of genotypic and environmental effects on biomass yield, nutritional and antinutritional factors in common vetch. Agronomy. 2022;**12**(7):1678

[17] Gangashetty PI, Yadav CB, Riyazaddin M, Vermula A, Asungre PA, Angarawai I, et al. Genotype-by-environment interactions for starch, mineral, and agronomic traits in pearl millet hybrids evaluated across five locations in West Africa. Frontiers in Plant Science. 2023;**23**(14):1171773

[18] Aguilar EG, Albarracín GD, Uñates MA, Piola HD, Camiña JM, Escudero NL. Evaluation of the nutritional quality of the grain protein of new amaranths varieties. Plant Foods for Human Nutrition. 2015;**70**:21-26

[19] Mallick SA, Azaz K, Gupta M, Sharma V, Sinha BK. Characterization of grain nutritional quality in wheat. Indian Journal of Plant Physiology. 2013;**18**:183-186

[20] AOAC. Official Methods of Analysis, Association of Official Analytical Chemists. 15th ed. USA: AOAC International; 1990

[21] Fassio A, Fernández EG, Restaino EA, La Manna A, Cozzolino D. Predicting the nutritive value of high moisture grain corn by near infrared reflectance spectroscopy. Computers and Electronics in Agriculture. 2009;**67**(1-2):59-63. DOI: 10.1016/j.compag.2009.03.001

[22] Nurit E, Tiessen A, Pixley KV, Palacios-Rojas N. Reliable and inexpensive colorimetric method for determining protein-bound tryptophan in maize kernels. Journal of Agricultural and Food Chemistry. 2009;**57**(16):7233-7238. DOI: 10.1021/jf901315x

[23] Rosales A, Galicia L, Oviedo E, Islas C, Palacios-Rojas N. Near-infrared reflectance spectroscopy (NIRS) for protein, tryptophan, and lysine evaluation in quality protein maize (QPM) breeding programs. Journal of Agricultural and Food Chemistry. 2011;**59**(20):10781-10786. DOI: 10.1021/jf201468x

[24] Porep JU, Kammerer DR, Carle R. On-line application of near infrared (NIR) spectroscopy in food production. Trends in Food Science and Technology. 2015;**46**(2):211-230. DOI: 10.1016/j.tifs.2015.10.002

[25] Williams P, El-Haramein FJ, Hani N, Safouh R. Crop quality evaluation methods and guidelines. Crop quality evaluation methods and guidelines. In: Technical Manual (ICARDA). International Center for Agricultural Research in the Dry Areas. 2nd ed. Vol. 14. Syria: Aleppo; 1988. p. 142

[26] Wu F, Rodricks JV. Forty years of food safety risk assessment: A history and analysis. Risk Analysis. 2020;**40**(S1):2218-2230

[27] Balkir P, Kemahlioglu K, Yucel U. Foodomics: A new approach in food quality and safety. Trends in Food Science and Technology. 2021;**108**:49-57

[28] Agelet LE, Hurburgh CR Jr. A tutorial on near infrared spectroscopy and its calibration. Critical Reviews in Analytical Chemistry. 2010;**40**:246-260. DOI: 10.1080/10408347.2010.515468

[29] Workman J, Weyer L. Practical Guide and Spectral Atlas for Interpretive

Near-Infrared. Boca Raton: CRC; 2012. DOI: 10.1201/b11894

[30] Villamuelas M, Serrano E, Espunyes J, Fernández N, López-Olvera JR, Garel M, et al. Predicting herbivore faecal nitrogen using a multispecies near-infrared reflectance spectroscopy calibration. PLoS One. 2017;**12**(4):e0176635. DOI: 10.1371/journal.pone.0176635

[31] Rukundo E. Stochastic model of rural agribusiness supply chain: A case study of Gatsibo District. Modern Economy. 2022;**13**:370-396. DOI: 10.4236/me.2022.133021

[32] Chadalavada K, Anbazhagan K, Ndour A, Choudhary S, Palmer W, Flynn JR, et al. NIR instruments and prediction methods for rapid access to grain protein content in multiple cereals. Sensors. 2022;**22**(10):3710. DOI: 10.3390/s22103710

[33] De Alencar Figueiredo LF, Davrieux F, Fliedel G, Rami JF, Chantereau J, Deu M, et al. Development of NIRS equations for food grain quality traits through exploitation of a core collection of cultivated sorghum. Journal of Agricultural and Food Chemistry. 2006;**54**(22):8501-8509. DOI: 10.1021/jf061054g

[34] Dos Santos CA, Lopo M, Páscoa RN, Lopes JA. A review on the applications of portable near-infrared spectrometers in the agro-food industry. Applied Spectroscopy. 2013;**67**(11):1215-1233

[35] Yang J, Li J, Hu J, Yang W, Zhang X, Xu J, et al. An interpretable deep learning approach for calibration transfer among multiple near-infrared instruments. Computers and Electronics in Agriculture. 2022;**192**:106584. DOI: 10.1016/j.compag.2021.106584

[36] Alamu EO, Nuwamanya E, Cornet D, Meghar K, Adesokan M, Tran T, et al. Near-infrared spectroscopy applications for high-throughput phenotyping for cassava and yam: A review. International Journal of Food Science and Technology. 2021;**56**(3):1491-1501

[37] Folli GS, Santos LP, Santos FD, Cunha PH, Schaffel IF, Borghi FT, et al. Food analysis by portable NIR spectrometer. Food Chemistry Advances. 2022;**1**:100074

[38] Beć KB, Grabska J, Siesler HW, Huck CW. Handheld near-infrared spectrometers: Where are we heading? NIR News. 2020;**31**(3-4):28-35

[39] Pasquini C. Near infrared spectroscopy: Fundamentals, practical aspects and analytical applications. Journal of the Brazilian Chemical Society. 2003;**14**:198-219

[40] Himmelsbach DS, Barton FE, McClung AM, Champagne ET. Protein and apparent amylose contents of milled rice by NIR-FT/Raman spectroscopy. Cereal Chemistry. 2001;**78**(4):488-492

[41] Kim KS, Park SH, Shim KB, Ryu SN. Use of near-infrared spectroscopy for estimating lignan glucosides contents in intact sesame seeds. Journal of Crop Science and Biotechnology. 2007;**10**(3):185-192. DOI: 10.1007/978-94-007-0723-8_13

[42] Elfadl E, Reinbrecht C, Claupeina W. Development of near infrared reflectance spectroscopy (NIRS) calibration model for estimation of oil content in a worldwide safflower germplasm collection. International Journal of Plant Production. 2012;**4**(4):259-270. DOI: 10.22069/IJPP.2012.709

[43] Beć KB, Grabska J, Huck CW. Principles and applications of miniaturized near-infrared (NIR) spectrometers. Chemistry–A European

Journal. 2021;**27**(5):1514-1532. DOI: 10.1002/chem.202002838

[44] Peruchi LC, Nunes LC, de Carvalho GG, Guerra MB, de Almeida E, Rufini IA, et al. Determination of inorganic nutrients in wheat flour by laser-induced breakdown spectroscopy and energy dispersive X-ray fluorescence spectrometry. Spectrochemical Acta Part B: Atomic Spectroscopy. 2014;**100**:129-136. DOI: 10.1016/j.sab.2014.08.025

[45] Fleming DE, Foran KA, Kim JS, Guernsey JR. Portable x-ray fluorescence for assessing trace elements in rice and rice products: Comparison with inductively coupled plasma-mass spectrometry. Applied Radiation and Isotopes. 2015;**104**:217-223. DOI: 10.1016/j.apradiso.2015.07.014

[46] Chen ZW, Gibson WM, Huang H. High definition X-ray fluorescence: Principles and techniques. X-Rays Optics and Instrumentation. 2008;**2008**:1-10. DOI: 10.1155/2008/318171

[47] Marguí E, Queralt I, De Almeida E. X-ray fluorescence spectrometry for environmental analysis: Basic principles, instrumentation, applications and recent trends. Chemosphere. 2022;**303**:135006. DOI: 10.1016/j.chemosphere.2022.135006

[48] Kak AC, Slaney M. Principles of computerized tomographic imaging. Society for Industrial and Applied Mathematics. 2001;**20**(23):114-133

[49] Hu W, Zhang C, Jiang Y, Huang C, Liu Q, Xiong L, et al. Nondestructive 3D image analysis pipeline to extract rice grain traits using X-ray computed tomography. Plant Phenomics. 2020;**2020**:1-12. DOI: 10.34133/2020/3414926

[50] Domhoefer M, Chakraborty D, Hufnagel E, Claußen J, Wörlein N, Voorhaar M, et al. X-ray driven peanut trait estimation: Computer vision aided agri-system transformation. Plant Methods. 2022;**18**(1):76. DOI: 10.1186/s13007-022-00909-8

[51] Long DA. Raman spectroscopy. New York. 1977;**1**:310

[52] Yang D, Ying Y. Applications of Raman spectroscopy in agricultural products and food analysis: A review. Applied Spectroscopy Reviews. 2011;**46**(7):539-560. DOI: 10.1080/05704928.2011.593216

[53] Yuan J, Sun C, Guo X, Yang T, Wang H, Fu S, et al. A rapid Raman detection of deoxynivalenol in agricultural products. Food Chemistry. 2017;**221**:797-802. DOI: 10.1016/j.foodchem.2016.11.101

[54] Krimmer M, Farber C, Kurouski D. Rapid and noninvasive typing and assessment of nutrient content of maize kernels using a handheld Raman spectrometer. ACS Omega. 2019;**4**(15):16330-16335. DOI: 10.1021/acsomega.9b01661

[55] Rai KN, Yadav OP, Govindaraj M, Pfeiffer WH, Yadav HP, Rajpurohit BS, et al. Grain iron and zinc densities in released and commercial cultivars of pearl millet (Pennisetum glaucum). Indian Journal of Agricultural Sciences. 2016;**86**(03):11-16

[56] Govindaraj M, Rai KN, Cherian B, Pfeiffer WH, Kanatti A, Shivade H. Breeding biofortified pearl millet varieties and hybrids to enhance millet Markets for Human Nutrition. Agriculture. 2019;**9**(5):106. DOI: 10.3390/agriculture9050106

[57] Fuchs T, Keßling P, Firsching M, Nachtrab F, Scholz G. Industrial applications of dual X-ray energy

computed tomography (2X-CT). In: Güneş O, Akkaya Y, editors. Nondestructive Testing of Materials and Structures. Vol. 6. Dordrecht: RILEM Bookseries, Springer; 2013. pp. 97-103. DOI: 10.1007/978-94-007-0723-8_13

[58] De Chiffre L, Carmignato S, Kruth JP, Schmitt R, Weckenmann A. Industrial applications of computed tomography. CIRP Annals. 2014;**63**(2):655-677. DOI: 10.1016/j.cirp.2014.05.011

[59] Walter A, Liebisch F, Hund A. Plant phenotyping: From bean weighing to image analysis. Plant Methods. 2015;**11**(14):1-1. DOI: 10.1186/s13007-015-0056-8

[60] Gomez FE, Carvalho G, Shi F, Muliana AH, Rooney WL. High throughput phenotyping of morpho-anatomical stem properties using X-ray computed tomography in sorghum. Plant Methods. 2018;**14**:1-3. DOI: 10.1186/s13007-018-0326-3

[61] Wu D, Guo Z, Ye J, Feng H, Liu J, Chen G, et al. Combining high-throughput micro-CT-RGB phenotyping and genome-wide association study to dissect the genetic architecture of tiller growth in rice. Journal of Experimental Botany. 2019;**70**(2):545-561. DOI: 10.1093/jxb/ery373

[62] Sohn M, Himmelsbach DS, Barton FE. A comparative study of Fourier transforms Raman and NIR spectroscopic methods for assessment of protein and apparent amylose in rice. Cereal Chemistry. 2004;**81**(4):429-433

[63] Ma CY, Lee PD. FT-Raman spectroscopy and its applications in cereal science. Cereal Chemistry. 2002;**79**(2):171-177. DOI: 10.1094/CCHEM.2002.79.2.171

[64] Payne WZ, Kurouski D. Raman-based diagnostics of biotic and abiotic stresses in plants. A Review. Frontiers in Plant Science. 2021;**11**:616672. DOI: 10.3389/fpls.2020.616672

[65] Saletnik A, Saletnik B, Puchalski C. Raman method in identification of species and varieties, assessment of plant maturity and crop quality—A review. Molecules. 2022;**27**(14):4454. DOI: 10.3390/molecules27144454

[66] Farber C, Kurouski D. Raman spectroscopy and machine learning for agricultural applications: Chemometric assessment of spectroscopic signatures of plants as the essential step toward digital farming. Frontiers in Plant Science. 2022;**13**:887511. DOI: 10.3389/fpls.2022.887511

[67] Gan Q, Wang X, Wang Y, Xie Z, Tian Y, Lu Y. Culture-free detection of crop pathogens at the single-cell level by micro-raman spectroscopy. Advanced Science. 2017;**4**(11):1700127. DOI: 10.1002/advs.201700127

[68] Kahriman F, Egesel CÖ. Development of a calibration model to estimate quality traits in wheat flour using NIR (near infrared reflectance) spectroscopy. Research Journal of Agricultural Science. 2011;**43**(3):392-400

[69] Dardenne P. Stability of NIR spectroscopy equations. NIR News. 1996;7(5):8-9. DOI: 10.1255/nirn.380

Chapter 7

Perspective Chapter: Advances in Modern Cereal Storage Systems

Zhichao Li, Pengfei Zhang and Qi Zhang

Abstract

Cereal storage is a critical link in the food supply chain, impacting food security, quality, and sustainability. The present chapter provides a comprehensive examination of cereal storage systems in terms of engineering design, manufacturing planning, construction management, technology solutions, and business functions, by focusing on four key facility elements: I. factory and plant layout; II. warehouse and transportation; III. automation, robotics, and sensors; and IV. postharvesting equipment. Through a systematic lens, engineering design principles, construction management strategies, and technological solutions, etc., are explored. It also highlights the interdependence of facility elements and their impact on business functions, emphasizing the need for well-planned layouts and integrated technologies. The chapter delves into advanced manufacturing processes, automation, and real-time monitoring as essential components for optimal cereal storage. Additionally, it introduces the concept of turnkey projects and one-stop services as modern business functions in this domain. Offering practical insights, it can serve as a valuable resource for professionals and enthusiasts seeking a holistic understanding of cereal storage systems, bridging the gap between theory and practical application.

Keywords: cereal storage systems, engineering design, facilities, one-stop service, turnkey projects

1. Introduction

Modern cereal storage systems represent a critical component of the agricultural supply chain, playing a pivotal role in ensuring food security, reducing postharvest losses, and meeting the demands of a growing global population. These systems encompass a range of innovative concepts and technologies designed to optimize the storage, preservation, and distribution of cereal crops such as wheat, rice, maize, and barley. They are essential for safeguarding food supplies, supporting agricultural sustainability, and promoting economic development in both rural and urban communities [1–3].

Viewing the cereal storage system as an exemplary production system allows us to conceptualize it as an integrated framework. This framework encompasses personnel, equipment, and procedures that work in concert to efficiently manage cereal storage operations. By adopting this perspective, we recognize the interconnectedness of various elements within the system and the importance of coordination and optimization to achieve optimal performance and outcomes. According to the definition of the production system [4], the cereal storage consists of two primary components: *facilities* and *storage support systems*, depicted in **Figure 1**.

As shown in **Figure 1**, the physical facilities of the cereal storage system further encompass distinct elements:

I. *Factory and Plant Layout*: The physical arrangement and design of the storage facility, including the layout of buildings, machinery, and operational zones. It involves optimizing spatial organization to enhance workflow efficiency and accessibility for storage and processing activities.

II. *Storage and Transportation*: The infrastructure dedicated to the physical storage and transportation of cereals, encompassing silos, bins, conveyors, and transport vehicles. This component emphasizes the hardware elements involved in maintaining the integrity of stored cereals and facilitating their movement within the facility.

III. *Automation, Robotics, and Sensors*: The incorporation of techniques such as automated systems, robotic devices, and sensor apparatuses. These elements are designed to enhance the efficiency of cereal storage operations by automating tasks, improving accuracy, and providing real-time data using physical hardware.

IV. *Postharvesting*: The tools and machinery employed in postharvest processes, including cleaning, drying, packaging, and screening equipment. This component focuses on the tangible devices and structures utilized in ensuring the quality and readiness of cereals for storage and distribution.

To ensure the efficient operation of cereal storage system facilities, a company must strategically organize itself to design processes and equipment, plan and control

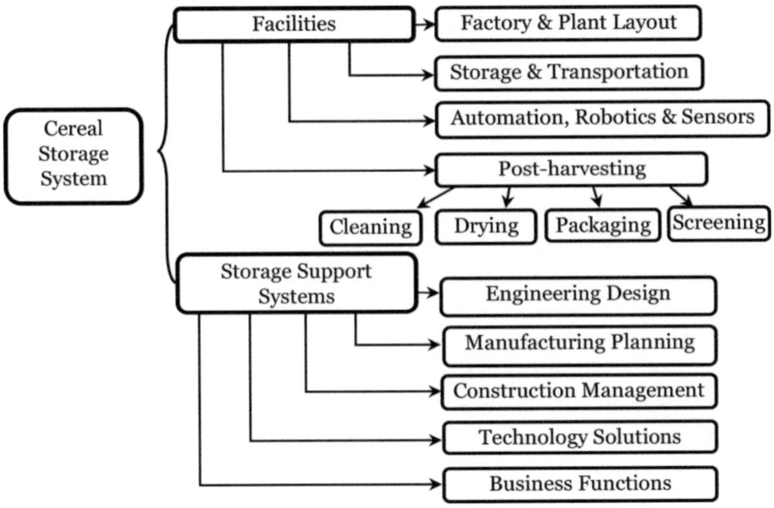

Figure 1.
Cereal storage system structure with components.

operations, and meet stringent quality requirements. These vital functions are executed through storage support systems—comprising both people and procedures, which form the backbone of how a company manages its storage operations. Many of these support systems do not have direct contact with the cereal itself; rather, they plan and control the cereal throughout the entire process.

The cereal storage support systems encompass a sequence of activities that exert widespread control or interaction with facilities and their components, as illustrated in **Figures 1** and **2**. These activities involve five key functions that necessitate extensive information flow and data processing:

1. Engineering design

2. Manufacturing planning

3. Construction management

4. Technology solutions

5. Business functions

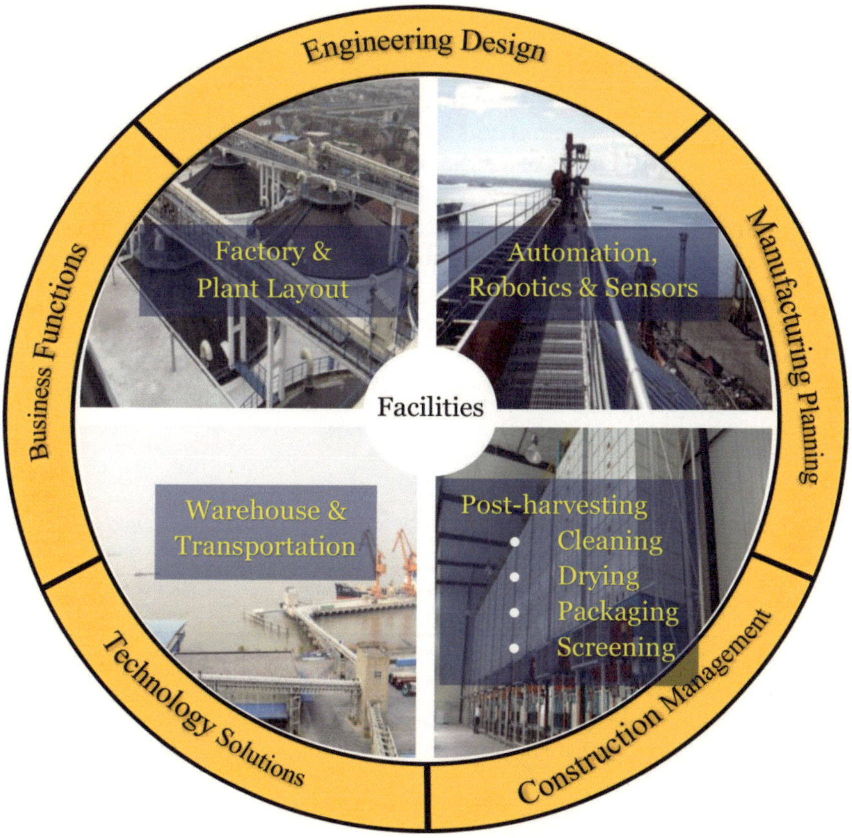

Figure 2.
Facilities and storage support systems. (photo courtesy of FAMSUM group Co., ltd.).

The present chapter will unravel the latest advancements and potential trends in cereal storage systems, focusing on the dynamic interplay between physical facilities and support systems. Covering factory and plant layout, storage and transportation infrastructure, automation, robotics and sensors, and postharvesting equipment, it explores the pivotal roles of engineering design, manufacturing planning, construction management, technology solutions, and business functions within these facilities. It is accessible to a broad audience by demystifying technology-driven solutions like software applications and automation and highlighting how these advancements enhance cereal storage efficiency and sustainability. Tailored for researchers and agriculture stakeholders, it offers clear insights into the collaborative contributions of physical structures and support systems, providing a concise and user-friendly guide to the evolving landscape of cereal storage systems.

2. Engineering design

In the context of cereal storage systems, engineering design can be defined as a structured set of decision-making processes and activities applied to ascertain the optimal form and configuration of storage facilities, components, etc. [5]. It extends beyond the physical structures to encompass the layout of facilities, the selection and design of equipment, and the integration of technologies, by taking into account factors such as spatial optimization, structural integrity, and workflow efficiency. It involves a dynamic interplay of conceptualization, planning, and execution, ensuring that the final form of the cereal storage system aligns seamlessly with the intended functions and addresses the complex demands of the cereal storage landscape. **Table 1** summarizes key engineering design principles, applications, and exemplary software/tools associated with the components of factory and plant layout; warehouse and transportation; automation, robotics, and sensors; as well as postharvesting equipment in the context of cereal storage systems.

For the cereal storage systems, the application of engineering design principles, methods, and technologies is the most important function in realizing and optimizing the efficiency and functionality of various facility components. When conceptualizing factory and plant layouts, engineers employ systematic planning to ensure

Facility components	Engineering design principles and applications	Exemplary software and tools
Factory and Plant Layout	Facility Design, Layout Workflow Design	Computer Aided Design (AutoCAD, Revit, etc.)
Warehouse and Transportation	Facility Design, Logistics Planning, Operation Research	Simulation Tools (AnyLogic, Simio, etc.)
Automation, Robotics and Sensors	Computer Integrated Manufacturing System(CIMS) Design, Sensor Design	Robotics Programming (Labview, PLC, etc.)
Postharvesting	Machine Design, Process Optimization Design	Finite Element Analysis (Solidworks, ANSYS, etc.)

Table 1.
Engineering design principles and applications and exemplary software and tools for cereal storage systems.

seamless processes, minimizing bottlenecks and maximizing space utilization [6]. Warehouse and transportation systems need to be designed with precision to streamline the movement of cereals, incorporating innovative storage solutions such as automated retrieval systems and conveyor belts to enhance operational speed and accuracy [7]. Automation, robotics, and sensor technologies play a crucial role in cereal storage, with robotic arms facilitating precise material handling, sensors ensuring real-time monitoring of environmental conditions, and automation streamlining repetitive tasks [8]. Furthermore, postharvesting equipment is designed and manufactured with a focus on gentle handling and preservation of cereal quality, employing technologies like drying systems and sorting machines to enhance efficiency while minimizing losses [9]. The integration of these engineering principles and advancements ensures a comprehensive and optimized cereal storage ecosystem, fostering sustainability and resilience in the agricultural supply chain.

Recent advances in the engineering design of facilities for cereal storage systems have been characterized by an increased integration of digital technologies, automation, and data-driven decision-making. Advancements and potential trends in the engineering design of various facility components for cereal storage systems based on historical patterns and common areas of innovation are summarized (but not limited to) below:

2.1 Factory and plant layout

- Flexible layouts to accommodate changing storage needs and product varieties [10]

- Modular and scalable design principles to allow for easy expansion or modification [11]

- Digital twins (DT) technology for virtual modeling and simulation of plant layouts for accurate design and operational predictions [12]

2.2 Warehouse and transportation

- Automated guided vehicles (AGVs) and autonomous mobile robots (AMRs) for efficient and flexible material handling in warehouses [13]

- Smart conveyor systems with integrated sensors for real-time monitoring of grain flow [14]

- Design optimization such as finite element analysis (FEA) of storage containers for configurative and structural optimization [15]

2.3 Automation, robotics, and sensors

- Robotics for tasks such as picking, packing, and stacking, improving the speed and accuracy of operations [16]

- Internet of Things (IoT) sensors and radio frequency identification (RFID) technology for continuous monitoring of environmental conditions, ensuring optimal storage conditions [17]

- Machine learning algorithms (MLA) for predictive maintenance, reducing downtime and enhancing the reliability of cereal storage systems [18]

2.4 Postharvesting equipment

- Energy-efficient (renewable energy sources incorporated) and environmentally benign drying systems to preserve grain quality [19]

- Optical sorting and computer vision technologies to enhance the precision of sorting processes and improve overall cereal grain quality [20]

- Intelligent packaging solutions with built-in sensors for real-time monitoring of product freshness and quality [21]

3. Manufacturing planning

Manufacturing is the process of converting raw materials, components, or parts into finished goods that meet a customer's expectations. Traditional manufacturing processes such as casting and roll forming have been playing a crucial role in producing the structural components for the cereal storage infrastructure like steel silos. Manufacturing planning encompasses the strategic selection of materials, manufacturing processes, and assembly techniques. **Table 2** summarizes manufacturing technologies and processes with exemplary applications (illustrated in **Figure 3**) in the cereal storage systems.

3D concrete printing is an innovative fast prototyping technology with possible potential to revolutionize factory and plant layout facility in the cereal storage systems [22]. This cutting-edge method involves layer-by-layer deposition of concrete, allowing for the rapid and cost-effective production of complex structures [26]. In cereal storage applications, 3D concrete printing offers advantages such as customizable and modular designs, enhanced structural integrity, and the ability to create intricately detailed configurations [27]. The technology could facilitate the construction of storage units with optimized geometries for efficient space utilization and tailored features for temperature control and aeration [26]. Furthermore, 3D concrete printing minimizes material waste and construction time, aligning with sustainability goals in agricultural infrastructure development [26, 27].

Facility components	Manufacturing technologies and processes	Exemplary applications
Factory and Plant Layout	Fast Prototyping Process	3D Concrete Printing and Modular Silos [19]
Warehouse and Transportation	Materials technologies, Advanced Manufacturing Processes	Roll Forming of Steel Silo Wall Parts [20]
Automation, Robotics and Sensors	Precision Machining and Assembly, Electronic Packaging	Robots for grain bin operation [21]
Postharvesting	Advanced Manufacturing Processes and Precision Assembly	Postharvesting Equipment Assembly [22]

Table 2.
Manufacturing technologies and processes and exemplary applications for cereal storage systems.

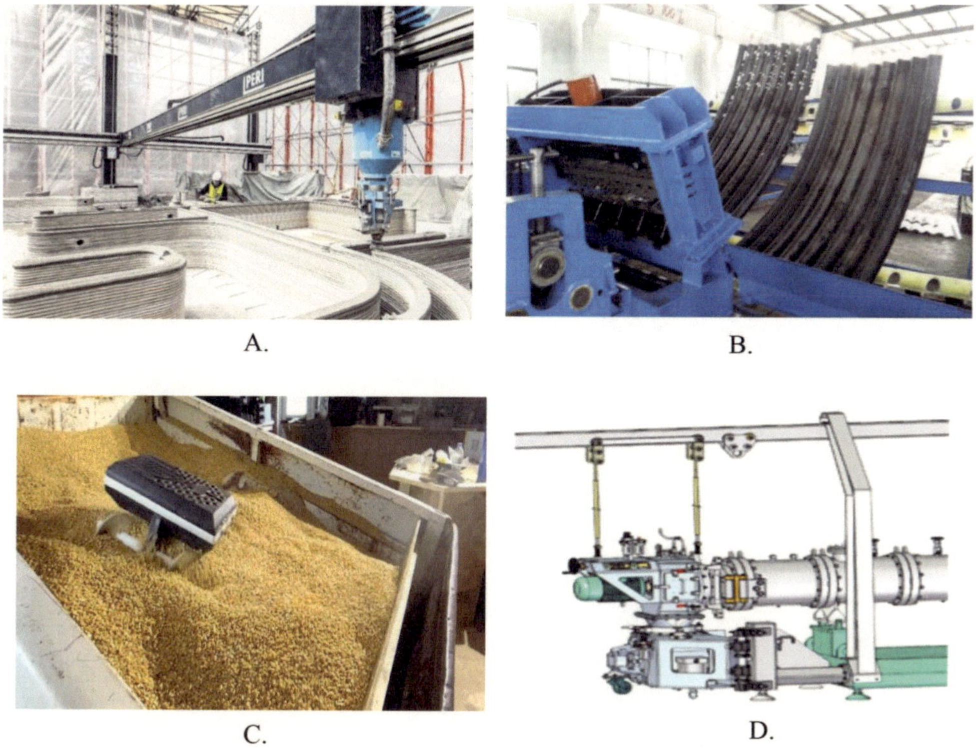

Figure 3.
Exemplary applications of manufacturing planning. A. 3D concrete printing for modular silos [22] (photo courtesy of Batchcrete international). B. Roll forming of steel Silo Wall parts [23] (photo courtesy of Rishbin Co., ltd.). C. Robot for grain barn Shoveling [24] (photo courtesy of FarmShow magazine). D. Assembly of pressure control system for post-harvesting equipment [25] (photo courtesy of FAMSUM group Co., ltd.).

In addition to innovative additive manufacturing processes like 3D concrete printing, the progress in manufacturing planning within the traditional paradigm of cereal storage systems can be examined concerning advancements in material technology, manufacturing processes, and precision assembly. **Figure 4** illustrates how typical manufacturing planning tackles key decisions such as choosing materials, picking the right manufacturing methods, and figuring out how to put parts together. It ensures that every step of making something is well prepared, from selecting materials that work well together to using efficient manufacturing techniques and assembling parts carefully. This thorough planning is crucial for making sure the final product is of good quality and durable.

3.1 Material technology

In the realm of manufacturing planning for cereal storage systems, the choice of materials plays a pivotal role in determining the longevity and performance of components. Advanced material technologies are harnessed to withstand the challenging environmental conditions and the demands of continuous operation. Selections such as corrosion-resistant alloys [28], high-strength composites [29, 30], and innovative coatings and paintings [31, 32] are strategically made to ensure durability and resilience against the varying conditions within storage facilities.

Figure 4.
Illustration of one-stop service (after [17]).

3.2 Manufacturing processes

Traditional manufacturing processes such as casting, forming, and machining lie at the heart of manufacturing planning for cereal storage systems. Leveraging advanced manufacturing processes, including precision casting, roll forming, and CNC machining, facilitates the creation of components with fine dimensions and tolerances. Precision casting involves highly controlled processes to create intricate and high-quality components [33]. In cereal storage, precision casting may be applied to produce critical elements with enhanced accuracy and reduced material waste. Roll forming is an advanced process where metal sheets are passed through sets of rolls to gradually shape them as shown in **Figure 3**. In cereal storage construction, roll forming is utilized to produce cylindrical sections for silos, ensuring uniformity and structural integrity [30, 34].

3.3 Precision assembly

The meticulous assembly of components is a cornerstone of manufacturing planning, where tight tolerances and seamless integration are paramount. Precision assembly techniques, guided by stringent quality control measures, guarantee that each component fits seamlessly into the broader cereal storage system. Robotics are increasingly employed in assembly processes, enhancing efficiency and reducing the margin for error [35]. They will be contributing to the overall reliability of the storage infrastructure in the future.

4. Construction management

Construction management is the strategic planning, coordination, and oversight of all phases in a construction project, including design, procurement, and execution [36]. The effective construction management in a cereal storage system considers not only the physical construction of facilities but also the integration of technology, automation, and sustainability principles. This means not only ensuring that the construction process is executed efficiently and to a high standard but also leveraging technology and automation to streamline operations, enhance productivity, and improve overall performance. Additionally, sustainability principles are integrated throughout the construction process to minimize environmental impact, optimize resource utilization, and promote long-term sustainability of the cereal storage system [37]. By considering these factors in construction management, stakeholders can ensure that the cereal storage system is not only well-built but also aligned with modern industry practices and environmental goals [37, 38].

4.1 Factory and plant layout

Construction Planning: The construction management process for factory and plant layout involves meticulous planning to optimize the arrangement of production units, machinery, and storage areas [39]. Attention should be given to create a layout that minimizes bottlenecks, allows for efficient material flow, and accommodates future expansion.

Modularity and Scalability: Construction teams focus on designing modular structures that can be easily expanded or reconfigured [40]. This modularity enhances flexibility and adaptability to changing storage needs, ensuring a cost-effective and scalable facility.

4.2 Warehouse and transportation

Structural Integrity: Construction management in warehouse and transportation systems emphasizes the construction of robust storage structures and efficient material handling systems. Attention is given to the structural integrity of warehouse components and the implementation of durable conveyor and transportation systems.

Safety and Accessibility: Construction teams prioritize safety measures, including the installation of proper signage, lighting, and safety barriers. Accessibility considerations are integrated into the construction of ramps, loading docks, and storage racks.

4.3 Automation, robotics, and sensor technologies

Infrastructure for Automation: Construction management for automation, robotics, and sensor technologies involves creating the necessary infrastructure to support these systems. This includes installing power and data lines, creating secure enclosures for robotic equipment, and ensuring connectivity for sensor networks.

Precision Construction: Attention to detail is crucial to ensure precision in the construction of components that house and support automated systems. The construction team works closely with technology specialists to integrate the physical infrastructure seamlessly with automation technologies.

4.4 Postharvesting equipment

Specialized Construction: Construction management for postharvesting equipment focuses on creating specialized structures to house equipment like sorting machines, dryers, and processing units. The emphasis is on precision in construction to accommodate the unique requirements of each piece of equipment.

Energy-Efficient Construction: With a growing emphasis on sustainability, construction management incorporates energy-efficient practices. This includes the installation of energy-efficient lighting, insulation, and the integration of renewable energy sources where feasible.

5. Technology solutions

In cereal storage systems, technology solutions are essential for efficiency and quality. These include but are not limited to automated cleaning, smart pest control, and precise drying technology. For instance, robotics help with sorting, ensuring only top-quality cereals are stored [41]. Technologies like radio-frequency identification (RFID) improve inventory tracking, making order fulfillment more efficient [42]. Innovative materials and manufacturing processes, like precision casting, create components with high precision [29–35]. Overall, these technology solutions streamline cereal storage, from production to distribution.

5.1 Factory and plant layout

Automated Cleaning Systems: Utilizing advanced cleaning technologies, such as automated vacuum systems and conveyor-based cleaning processes, within the factory layout ensures efficient removal of debris and contaminants, maintaining a clean production environment [43].

Intelligent Pest Control: Traditional pest control methods for cereal storage typically involve physical, chemical, and biological approaches [44]. Physical means maintaining cleanliness, ventilation, and sealing storage facilities to minimize pest entry. Chemical ways involve the application of insecticides to control insect populations, while biological methods utilize natural predators or parasites to target pest species. These methods have been used for decades and remain integral to factory and plant layout for modern cereal storage systems [44]. Nowadays, intelligent pest control technologies, including automated traps and monitoring systems, are incorporated into the factory layout to detect and address pest issues promptly, safeguarding stored cereals [44].

5.2 Warehouse and transportation systems

Automated Material Handling: Automated conveyor and robotic systems are employed in warehouse layouts for efficient material handling [43]. These technologies streamline the movement of cereals from storage to transportation points, minimizing manual labor and reducing the risk of damage.

Real-time Inventory Tracking: Warehouse and transportation systems integrate real-time monitoring solutions, such as RFID or barcode technologies, to provide accurate and up-to-date inventory tracking [42]. This ensures precise control over stock levels and supports efficient order fulfillment.

5.3 Automation, robotics, and sensor technologies

Automated Sorting Systems: Robotics and sensor technologies are applied for automated sorting of cereals based on quality, size, and other parameters. This enhances the efficiency of postharvest processes and ensures that only high-quality grains are stored and distributed [45].

Real-time Environmental Monitoring: Sensors for temperature, humidity, and gas levels are strategically placed throughout the facility to provide real-time data, enabling continuous monitoring of storage conditions and allowing for immediate corrective actions if deviations occur [17].

5.4 Postharvesting equipment

Precision Drying Technologies: Advanced drying systems, such as infrared [46] and microwave [47] drying, are integrated into postharvesting equipment. These technologies ensure precise control over the drying process, preventing over-drying or moisture retention, which can impact the quality of stored cereals.

Smart Sorting and Grading: Postharvest equipment incorporates smart sorting and grading technologies, utilizing machine vision and artificial intelligence to assess and categorize cereals based on quality parameters [45]. This enhances the accuracy and speed of the sorting process.

Aforementioned technology solutions collectively contribute to the optimization of cereal storage systems by enhancing efficiency, ensuring quality control, and enabling real-time monitoring for proactive management of storage conditions. The integration of automation, robotics, and sensor technologies across the various facility elements represents a holistic approach to modernizing cereal storage processes.

6. Business functions

Business functions for cereal storage systems rely on well-designed facility elements. A smart factory layout, effective warehouse systems, integrated automation, and advanced postharvest equipment contribute to smooth operations and better quality control. Summing up these functions, the approach can be seen as a "Turnkey Project" and a "One-stop Service," showcasing a comprehensive and simplified strategy for cereal storage efficiency.

6.1 Turnkey project

The "turnkey" concept represents a comprehensive approach where a contractor or a project team takes full responsibility for the entire lifecycle of the storage facility, from design and construction to commissioning and handover [48, 49]. This approach is increasingly applied in the advancement of cereal storage systems to provide clients with a seamless and hassle-free experience. **Table 3** shows the turnkey approach advancements in cereal storage compared with the traditional approach.

	Traditional approach	Turnkey approach
Design and Engineering	Clients separately engage architects, engineers, and other specialists.	Integrated design and engineering services. A specialized team is responsible for creating a comprehensive design that aligns with the client's requirements and industry standards.
Construction and Implementation	Clients coordinates with multiple contractors for different aspects of construction, such as structural work, equipment installation, and automation systems.	Streamlines construction by entrusting a single entity with the entire construction process. This includes site preparation, structural construction, installation of storage equipment, and integration of automation and monitoring systems.
Technology Integration	Integrating technologies like automation, robotics, and sensor systems require coordination among various vendors and specialists.	Leverage advancements in technology integration. The project team ensures seamless incorporation of the latest technologies for efficient material handling, real-time monitoring, and automated processes within the cereal storage facility.
Quality Assurance and Compliance	Ensuring compliance with industry regulations and quality standards involve separate assessments and inspections.	Prioritize quality assurance and compliance from the outset. The integrated team manages and oversees the entire process, ensuring that the facility meets or exceeds relevant regulations and quality benchmarks.
Operational Handover	Clients may need to coordinate the transition from construction to operations, involving multiple handovers and potential operational disruptions.	Ensures a smooth transition from construction to operations. The facility is handed over in a fully operational state, minimizing downtime and allowing clients to immediately use the cereal storage system.
Lifecycle Management	Clients may need to independently manage ongoing maintenance and upgrades.	Include ongoing maintenance and lifecycle management services. The contractor takes responsibility for ensuring the ongoing efficiency and functionality of the cereal storage system.

Table 3.
Comparison of traditional approach vs. turnkey approach [45].

In summary, the turnkey concept in cereal storage system advancements provides clients with an end-to-end solution, simplifying the entire process from conception to operation. This integrated approach leverages advancements in design, construction, technology integration, and ongoing management to deliver efficient, high-quality, and fully operational cereal storage facilities.

6.2 One-stop service

In the development of cereal storage systems, "One-stop Service" means getting all services needed for designing, building, and managing the storage facility from a single provider [50]. This includes everything from planning and design to construction, logistics, installation, and training. It includes various stages such as consultation, design, civil construction, manufacture, logistics, installation and commissioning, training, service, etc., as shown in **Figure 4**.

For the consultation stage, a dedicated team of experts typically from the company's consulting department will handle the process. These professionals are well-versed in conducting due diligence and investment evaluations, providing technical consultations, navigating approval procedures, negotiating contracts, estimating project costs, assessing risks, and offering ongoing support. They work closely with clients to understand their specific needs and objectives, guiding them through every step of the project planning and decision-making process. This collaborative approach ensures that clients receive tailored solutions that align with their goals while maximizing operational efficiency and minimizing risks.

During the design stage, the emphasis is on creating customer-centric solutions. This involves carefully analyzing market trends, evaluating customer needs, and assessing feasibility to tailor existing products and develop new solutions. By leveraging market analysis and customer feedback, the design team identifies opportunities to optimize existing offerings and explore new business avenues. Their goal is to unlock new potential by adapting to terrain specifics and delivering innovative solutions that address evolving market demands. Through a combination of market insight, customer input, and feasibility assessment, the design stage aims to deliver customized solutions that meet the unique requirements of each client while also driving business growth and expansion.

The civil construction phase focuses on enhancing construction processes to achieve improved safety, efficiency, and cost-effectiveness. The primary objectives include minimizing construction timelines, refining design elements, lowering expenses, and maximizing project benefits. By streamlining construction methods and optimizing project management practices, this phase aims to deliver superior outcomes while ensuring adherence to budgetary constraints and safety standards. Through meticulous planning and execution, the civil construction step strives to enhance project performance, reduce risks, and enhance overall project value.

The manufacturing stage is dedicated to ensuring the production of high precision, quality products through material processing, machining, painting, and assembly. This is achieved by leveraging advanced manufacturing technologies, including CNC machining centers, welding robots, streamlined production systems, and lean manufacturing principles. Each step of the manufacturing process is carefully executed to maintain strict quality standards and consistency. By employing cutting-edge technologies and efficient production methods, this stage aims to optimize product accuracy, minimize defects, and enhance overall product performance. Through a combination of advanced machinery and rigorous quality control measures, the manufacturing stage is instrumental in delivering reliable and top-quality products to meet customer expectations.

The logistics stage is focused on delivering efficient and reliable logistics solutions supported by modern technology. By utilizing advanced supply chain logistics information systems and management software, the company ensures seamless coordination and management of transportation processes. Additionally, specialized packaging designs incorporating barcode technology enhance inventory tracking and facilitate accurate cargo management. Through these measures, the company is committed to providing fast, precise, and dependable logistics services while prioritizing the safety and security of all shipments. This integration of cutting-edge technology and strategic logistics planning ensures timely delivery and customer satisfaction.

The installation and commissioning stage is dedicated to ensuring the successful implementation of projects by leveraging insights from past experiences to drive ongoing enhancements. Through continuous learning and improvement initiatives, the company strives to refine installation and commissioning processes, addressing any challenges encountered and optimizing project delivery. By drawing upon lessons learned from previous projects, the team identifies areas for improvement, implements best practices, and adopts innovative approaches to enhance efficiency and effectiveness. This commitment to continuous improvement fosters a culture of excellence and ensures that each project is executed flawlessly, meeting or exceeding client expectations.

The training stage aims to facilitate knowledge-sharing and create positive learning experiences for clients. The company organizes specialized training sessions led by experts in technology and management, covering various aspects such as production, operation, management, and maintenance. These training programs are designed to empower clients with the necessary skills and expertise to effectively utilize and maintain their equipment. By fostering professionals in these fields, the company not only addresses equipment-related challenges but also promotes collaboration and knowledge exchange among producers. Through these training initiatives, clients gain valuable insights and practical skills while also benefiting from networking opportunities and shared learning experiences within the industry.

The service stage represents a crucial aspect of the company's operations, focusing on delivering exceptional customer service to enhance the efficiency and value of customers' equipment and operations. Central to this objective is the expansion of service capabilities and support worldwide, ensuring accessibility and responsiveness to customer needs. Service engineers undergo continuous training, equipped with updated tools and support programs to stay abreast of the latest standards and technologies. A comprehensive service network is established, comprising local offices, spare part warehouses, field service teams, and online support, supplemented by remote services for prompt and efficient troubleshooting.

In essence, the company's customer service offerings encompass various key components:

- Spare parts service: Ensuring plant availability through timely access to essential components

- Repair service: Prioritizing reliability and minimizing downtime through swift and effective repairs

- Maintenance service: Proactively addressing issues before they arise, optimizing equipment performance

- Retrofit and reconditioning service: Breathing new life into older equipment, extending its operational lifespan

- Training and courses service: Providing enriching learning experiences for customers to enhance their skills and knowledge

- Process optimization service: Driving productivity gains and ensuring readiness to tackle evolving challenges

- Other services: Offering consulting, installation, commissioning, and other tailored solutions to meet diverse customer needs

By offering a comprehensive suite of service capabilities, the company aims to foster long-term partnerships with customers, supporting their success and driving mutual growth.

7. Concluding remarks

In conclusion, a systematic exploration of cereal storage systems is provided to unravel the intricate dynamics of facility elements crucial for their efficient functioning. Examining the perspectives of factory and plant layout, warehouse and transportation systems, automation, robotics, sensors, and postharvesting equipment, the chapter underscores the integrated nature of these components. From the lens of engineering design principles to the execution strategies of construction management, the symbiotic relationship between these facility elements and their profound impact on business functions becomes evident, setting the stage for effective cereal storage.

Technological solutions, ranging from precision manufacturing processes to real-time monitoring, showcase the ongoing evolution of the field. The introduction of turnkey projects and one-stop services exemplifies modern approaches to meet the complex demands of cereal storage systematically. The insights presented here will pave the way for future advancements and innovations in the ever-evolving landscape of cereal storage systems.

Acknowledgements

Special thanks to the FAMSUN Group Co., Ltd., Research and Development Institute for their valuable contributions and insights, enriching the content of this book chapter on cereal storage systems. Their expertise and dedication have played a pivotal role in shaping the discussions on facility elements, engineering design, and technological advancements within the field. Their ongoing commitment to excellence and innovation has significantly enhanced the depth and quality of this work.

Author details

Zhichao Li[1*], Pengfei Zhang[2] and Qi Zhang[3]

1 North Carolina A&T State University, Greensboro, USA

2 FAMSUN Group Co., Ltd., Yangzhou, China

3 Yangzhou University, Yangzhou, China

*Address all correspondence to: zli@ncat.edu

References

[1] Rosentrater KA. Chapter 3 - overview of storage systems for cereal grains and grain products. In: Rosentrater KA, editor. Storage of Cereal Grains and their Products. 5th ed. UK: Woodhead; 2022. DOI: 10.1016/B978-0-12-812758-2.00013-1

[2] Bala BK. Drying and Storage of Cereal Grains. 2nd ed. London: Wiley; 2016. ISBN: 978-1-119-12423-8

[3] Olorunfemi BJ, Kayode SE. Post-harvest loss and grain storage technology- a review. Turkish Journal of Agriculture - Food Science and Technology. 2021;**9**:75-83. DOI: 10.24925/turjaf.v9i1.75-83.3714

[4] Groover MP. Automation, Production Systems and Computer-Integrated Manufacturing. 5th ed. New York: Pearson; 2019. ISBN: 0133499618

[5] Rudolph J. Engineering Design. UK: Pearson Prentice Hall; 2005

[6] Tompkins JA, White JA. Facilities Planning. 4th ed. India: Wiley; 2010. ISBN: 978-0-470-44404-7

[7] Christopher M. Logistics & Supply Chain Management. 6th ed. London: Pearson; 2022. ISBN-13: 9781292416182

[8] Craig J. Introduction to Robotics. Global ed. New Jersey: Pearson; 2021. ISBN-13: 9781292164953

[9] Friedmann M, Maier D. Chapter 5 advances in bulk storage of cereals and grains. In: Maier D, editor. Advances in Postharvest Management of Cereals and Grains. 1st ed. London: Burleigh Dodds Science Publishing. DOI: 10.19103/AS.2020.0072

[10] Williams G, Rosentrater K. Design and engineering considerations for grain storage, handling, and processing facilities. In: Storage of Cereal Grains and their Products. 5th ed. UK: Woodhead; 2022. DOI: 10.1016/C2016-0-03912-8

[11] Custodio L, Machado R. Flexible automated warehouse: A literature review and an innovative framework. International Journal of Advanced Manufacturing Technology. 2020;**106**:533-558. DOI: 10.1007/s00170-019-04588-z

[12] Dyck G, Hawley E, Hildebrand K, Paliwal J. Digital twins: A novel traceability concept for post-harvest handling. Smart Agricultural Technology. 2023;**3**. DOI: 10.1016/j.atech.2022.100079

[13] Zhang J et al. Automated guided vehicles and autonomous mobile robots for recognition and tracking in civil engineering. Automation in Construction. 2023;**146**. DOI: 10.1016/j.autcon.2022.104699

[14] Maier G et al. Experimental evaluation of a novel sensor-based sorting approach featuring predictive real-time multi-object tracking. IEEE Transactions on Industrial Electronics. 2021;**68**:1548-1559. DOI: 10.1109/TIE.2020.2970643

[15] Khalil M, Ruggieri S, Uva G. Assessment of structural behavior, vulnerability, and risk of industrial silos: State-of-the-art and recent research trends. Applied Sciences. 2022;**12**(6):3006. DOI: 10.3390/app12063006

[16] Thakur A, Venu S, Gurusamy M. An extensive review on agricultural robots with a focus on their perception systems. Computers and Electronics in

Agriculture. 2023;**212**. DOI: 10.1016/j.compag.2023.108146

[17] Lutz É, Coradi PC. Applications of new technologies for monitoring and predicting grains quality stored: Sensors, internet of things, and artificial intelligence. Measurement. 2022;**188**. DOI: 10.1016/j.measurement.2021.110609

[18] Waldamichael FG et al. Machine learning in cereal crops disease detection: A review. Algorithms. 2022;**15**(3):75. DOI: 10.3390/a15030075

[19] Menon A et al. A systematic review on the recent advances of the energy efficiency improvements in non-conventional food drying technologies. Trends in Food Science & Technology. 2020;**100**:67-76. DOI: 10.1016/j.tifs.2020.03.014

[20] Zareef M et al. Recent advances in assessing qualitative and quantitative aspects of cereals using nondestructive techniques: A review. Trends in Food Science & Technology. 2021;**116**:815-828. DOI: 10.1016/j.tifs.2021.08.012

[21] Osmólska E, Stoma M, Starek-Wójcicka A. Application of biosensors, sensors, and tags in intelligent packaging used for food products—A review. Sensors. 2022;**22**(24):9956. DOI: 10.3390/s22249956

[22] Batchcrete International. An Overview of 3D Concrete Printing & Modular Silos [Internet]. 2022. Available from: https://batchcrete.net.au/an-overview-of-3d-concrete-printing-modular-silos/ [Accessed: January 12, 2024]

[23] Roll Form Lines Specialist. Grain Silo Roll Forming Machine [Internet]. 2020. Available from: http://www.rishbin.com/products/grain-silo-roll-forming-machine.html [Accessed: January 12, 2024]

[24] Anonymous. Grain bin robot designed to eliminate trips into bins. Farm Show Magazine [Internet]. 2021;**45**(2):23. Available from: https://www.farmshow.com/a_article.php?aid=36880 [Accessed: January 12, 2024]

[25] My View Integrated Solution Provider. FAMSUN: Enabling Resilience in Manufacturing, Services and Supply System [Internet]. 2021. Available from: http://enmp.famsungroup.com/Public/Upload/file/20210903/1630649463bc23d85073e89f58.pdf [Accessed: January 12, 2024]

[26] Mohan MK, Rahul AV, De Schutter G, Van Tittelboom K. Extrusion-based concrete 3D printing from a material perspective: A state-of-the-art review. Cement and Concrete Composites. 2021;**2021**:115. DOI: 10.1016/j.cemconcomp.2020.103855

[27] Fullard LA, Godfrey AJR, Manaf MF, Davies CE, Cliff A, Fukuoka M. Mixing experiments in 3D-printed silos: The role of wall friction and flow correcting inserts. Advanced Powder Technology. 2020;**31**:1915-1923. DOI: 10.1016/j.apt.2020.02.024

[28] Oki M, Anawe PAL. A review of corrosion in agricultural industries. Physical Science International Journal. 2015;**5**:216-222. DOI: 10.9734/PSIJ/2015/14847

[29] Ghadhban MY. Using composite material in transportation pipes of wheat in silos. Journal of Environmental Studies. 2010;**5**:1-10. DOI: 10.21608/jesj.2010.185295

[30] Gascons M, Blanco N, Matthys K. Evolution of manufacturing processes

for fiber-reinforced thermoset tanks, vessels, and silos: A review. IIE Transactions. 2012;**44**:476-489. DOI: 10.1080/0740817X.2011.590177

[31] Adesina J, Mobolade NB, Dinabandhu S, Yallappa R. Traditional methods of food grains preservation and storage in Nigeria and India. Annals of Agricultural Sciences. 2019;**64**:196-205. DOI: 10.1016/j.aoas.2019.12.003

[32] Cai Y et al. Performance of a superamphiphobic self-cleaning passive subambient daytime radiative cooling coating on grain and oil storage structures. Heliyon. 2023;**9**(4):e14599. DOI: 10.1016/j.heliyon.2023.e14599

[33] Singh S, Singh R. Precision investment casting: A state of art review and future trends. Proceedings of the Institution of Mechanical Engineers, Part B: Journal of Engineering Manufacture. 2016;**230**(12):2143-2164. DOI: 10.1177/0954405415597844

[34] Michael Rotter PJ. Metal silos, shells and containment structures. Progress in Structural Engineering and Materials. 2005;**1**(4):428-435. DOI: 10.1002/pse.2260010412

[35] Tehrani BM, BuHamdan S, Alwisy A. Robotics in assembly-based industrialized construction: A narrative review and a look forward. International Journal of Intelligent Robotics and Applications. 2023; 7:556-574. DOI: 10.1007/s41315-022-00257-9

[36] Harris F, McCaffer R, Baldwin A, Edum-Fotwe F. Modern Construction Management. 8th ed. UK: Wiley; 2021. p. 2021

[37] Stanitsas M, Kirytopoulos K, Leopoulos V. Integrating sustainability indicators into project management: The case of construction industry. Journal of Cleaner Production. 2021;**279**. DOI: 10.1016/j.jclepro.2020.123774

[38] McGeorge D, Zou P. Construction Management: New Directions. 3rd ed. London: Wiley; 2012. ISBN: 9780470674017

[39] Amer F, Koh HY, Golparvar-Fard M. Automated methods and Systems for Construction Planning and Scheduling: Critical review of three decades of research. Journal of Construction Engineering and Management. 2021;**47**:7. DOI: 10.1061/(ASCE) CO.1943-7862.0002093

[40] Islam M, Maxwell D, Bunster V, Couper R. Modularity as an enabler of scalability in industrialised building platforms. In: Kreutzberg A, editor. Envisioning Architectural Scales in the Analogue and Virtual Representation of Architecture: Proceedings of the 16th EAEA Conference. Denmark: Royal Danish Academy - Architecture, Design, Conservation; 2023. pp. 186-198. DOI: 10.60558/eaea16-2023-230

[41] Panda BK, Panigrahi SS, Mishra G, Kumar V. Robotics for general material handling machines in food plants. In: Transporting Operations of Food Materials within Food Factories. UK: Woodhead Publishing; 2023. pp. 341-372. DOI: 10.1016/B978-0-12-818585-8.00005-2

[42] Bibi F, Guillaume C, Gontard N, Sorli B. A review: RFID technology having sensing aptitudes for food industry and their contribution to tracking and monitoring of food products. Trends in Food Science & Technology. 2017;**62**:91-103. DOI: 10.1016/j.tifs.2017.01.013

[43] Caldwell DG, Davis ST, Masey RJM, Gray JO. Automation in food processing. In: Springer Handbook

of Automation. Berlin Heidelberg: Springer; 2009. pp. 1041-1059. DOI: 10.1007/978-3-540-78831-7_60

[44] Hamel D, Rozman V, Liška A. Storage of cereals in warehouses with or without pesticides. Insects. 2020;**11**(12):846. DOI: 10.3390/insects11120846

[45] Mishra G et al. Research advancements in optical imaging and spectroscopic techniques for nondestructive detection of mold infection and mycotoxins in cereal grains and nuts. Comprehensive Reviews in Food Science and Food Safety. 2021;**20**(5):4612-4651. DOI: 10.1111/1541-4337.12801

[46] Sakare P, Prasad N, Thombare N. Infrared drying of food materials: Recent advances. Food Engineering Reviews. 2020;**12**:381-398. DOI: 10.1007/s12393-020-09237-w

[47] Khodifad BC, Dhamsaniya NK. Drying of food materials by microwave energy-a review. International Journal of Current Microbiology and Applied Sciences. 2020;**9**:1950-1973. DOI: 10.20546/ijcmas.2020.905.223

[48] Jones KG, Mulder F, Morga M, Wanigarathna N. Improving organizational resilience: The TURNKey project. In: IOP Conference Series: Earth and Environmental Science CIB W070 Conference on Facility Management and Maintenance; 8-11 May 2023; Trondheim, Norway. Norway: Open Access-IOP Publishing; 2023. DOI: 10.1088/1755-1315/1176/1/012043

[49] Mshali RM. An Investigation into Construction Contracts in Malawi-Turnkey Versus Traditional Contracts [Doctoral Dissertation]. Bolton, England: Semantic Scholar: University of Bolton; 2016

[50] Delivery System Global Service [Internet]. 2022. Available from: http://en.famsungroup.com/service?tab=0 [Accessed: January 24, 2024]